IMAGES
of America

NOME

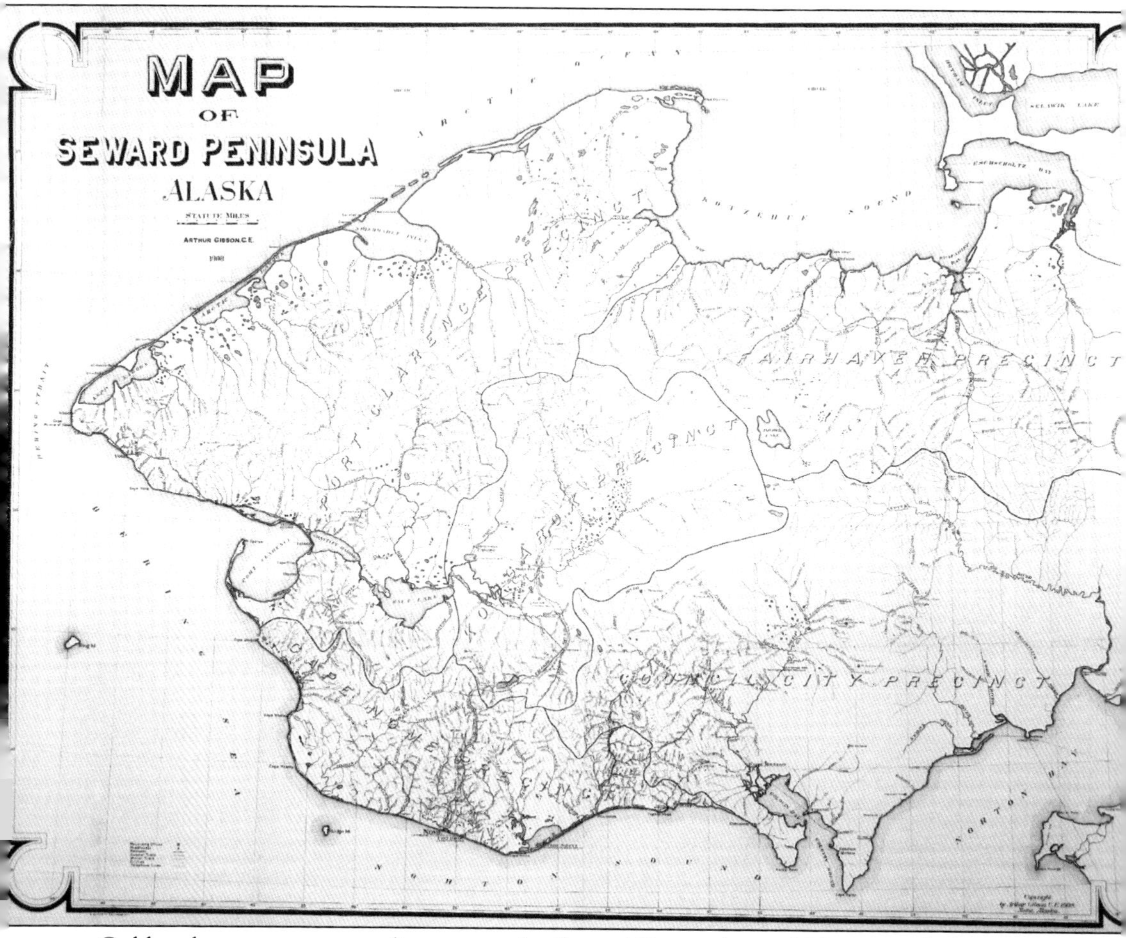

Gold rush prospectors traveling across the Seward Peninsula would have regularly consulted this map drawn by Arthur Gibson of the Nome Weather Bureau in 1908. The map includes the locations of significant roadhouses, railroads, telephone lines, and mining ditches. Summer and winter trails are marked, as well as US mail cabins for those navigating the rolling tundra by foot or dog team. (Carrie M. McLain Memorial Museum, 1995.1.55.)

ON THE COVER: Teams of sled dogs prepare to transport crates of supplies from the SS *Corwin* over seven miles of sea ice to Nome on June 8, 1911. The *Corwin* began her career in 1877 as a US revenue cutter that cruised the Bering Sea and Arctic Ocean for 23 years in government service. She was sold in 1900 and refitted as a merchant and charter vessel just in time to carry thousands of prospectors to Nome. (Photograph by H.G. Kaiser; Carrie M. McLain Memorial Museum, 2013.1.2.)

IMAGES
of America

NOME

Amy Phillips-Chan
Foreword by Cussy Kauer

ARCADIA
PUBLISHING

ISBN 978-1-4671-0291-9

Published by Arcadia Publishing
Charleston, South Carolina

Printed in the United States of America

Library of Congress Control Number: 2018953528

For all general information, please contact Arcadia Publishing:
Telephone 843-853-2070
Fax 843-853-0044
E-mail sales@arcadiapublishing.com
For customer service and orders:
Toll-Free 1-888-313-2665

Visit us on the Internet at www.arcadiapublishing.com

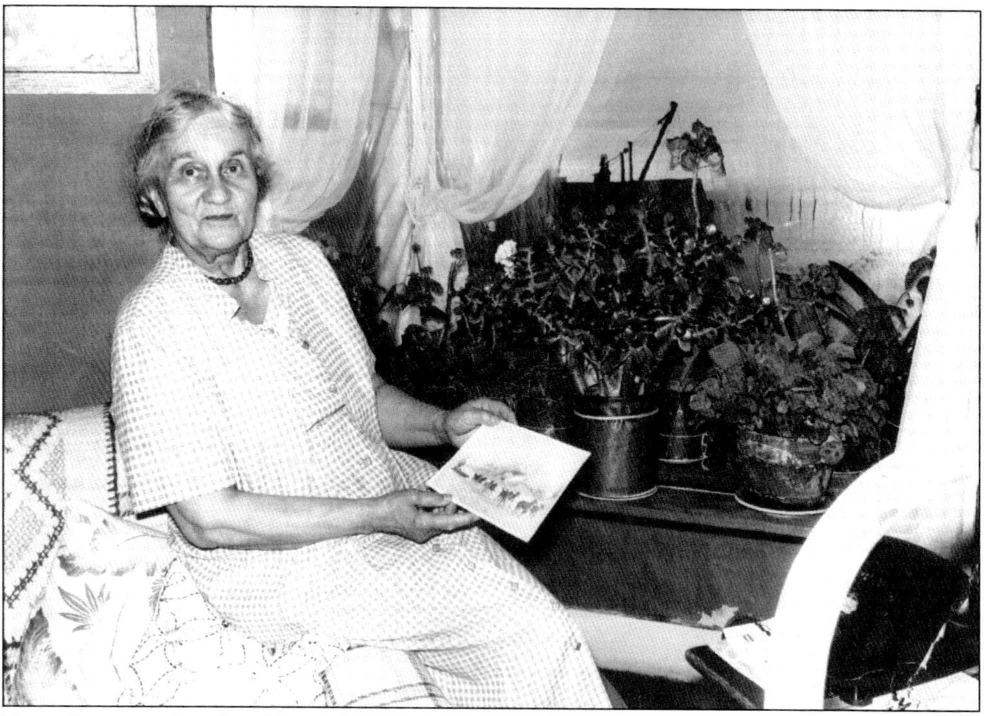

Nome historian Caroline "Carrie" Mary Stipek (1895–1973) arrived in Nome on the fringe of the gold rush in 1905. Upon retirement as Nome's city clerk, she embarked on an ambitious pursuit to acquire historical photographs and artifacts that helped establish the foundation for a museum in Nome. Carrie McLain is seen here in her home on Belmont Point around 1965, sharing a photograph from the All Alaska Sweepstakes sled dog race, which ran from 1908 to 1917. (Carrie M. McLain Memorial Museum, NMP-82-8-3.)

Contents

Foreword		6
Acknowledgments		7
Introduction		8
1.	Seasonal Cycle	9
2.	Tent City	23
3.	Gold Mining	31
4.	Ivory Carving	43
5.	Building a Town	49
6.	Nome Merchants	65
7.	Resourceful Home Builders	79
8.	Staying Connected	87
9.	Dogs, Reindeer, and Horses	99
10.	Wings and War	117
Bibliography		127

Foreword

My name is Caroline "Cussy" Kauer and I am a granddaughter of Carrie M. McLain. I was born in Nome and have lived here for almost 67 years, about the same length of time as Grandma Carrie. Both of us share a deep love of this community and this country, and have been strong promoters and supporters of Nome our adult lives.

After my grandmother passed in 1973, the City of Nome renamed the museum, in fitting tribute and in her honor, the Carrie M. McLain Memorial Museum. For over 20 years, Carrie had been reaching out and writing to friends who no longer lived in Nome with requests for donations to her collection of photographs and memorabilia in the hope of establishing a museum.

Carrie McLain was extremely proud of Nome and the territory. At every opportunity, she would box up her pictures and make presentations to visitors in her church, the chamber of commerce, and schools. She enjoyed sharing her historical knowledge when folks called at her home on Belmont Point. There, in her kitchen, Grandma would sit at her typewriter and compose many of the same notes and bits of information that I have used in my research. The list of information she left is long and includes names of people, events, and buildings—her fabulous history of Nome.

In this book, you will experience Nome through the eyes of Carrie McLain and the people she knew during the gold rush days. Steamships, tents, and dog sleds dominated the Victorian-looking town that had risen up from the beach sands and that she called home. Grandma Carrie was standing one day in her small living room, looking across the top of her geraniums in a little bay window toward town, and said something like, "Carol, the Nome I knew is now only in my mind's eye." I think I saw a tear coming down her cheek and I didn't know what she meant. I get it now, Grandma. I get it.

The Nome I knew growing up is fast becoming only in my mind's eye. Look at us now! Running water! Flushing toilets! Warm houses! Television! Computers! Internet! Facebook!

I can only hope we instill a deep appreciation for life as it was and recognition of the struggle and work and the creativity and ingenuity of those who came before us, who built Nome into what we know today. Thank you for sharing and enjoying our publication.

—Cussy Kauer

ACKNOWLEDGMENTS

Over 50 years ago, Carrie M. McLain had the foresight and determination that the visual and artifactual history of Nome should be preserved and shared for future generations. Her founding collection of photographs, letters, and objects has continued to encourage donations from individuals and families looking to return their treasures and memories to Nome. The Carrie M. McLain Memorial Museum now holds over 12,000 original prints and negatives documenting over 100 years of Nome history. This book is dedicated to all those local historians who have sought to visually capture the people and places that make Nome a home.

I came onboard as the director of the Carrie M. McLain Memorial Museum in 2015 to lead the development of a new museum. Three whirlwind years included taking inventory and cataloguing thousands of objects and photographs, laying out new museum spaces, designing storage areas, moving the museum collections, and developing new immersive exhibits. One of the most exciting aspects of the project involved opening boxes and discovering collections of photographs that offered glimpses into the daily lives of Nome residents. The vast majority of these images have not been shared with the public until now.

My sincerest appreciation to the community members of Nome who have shared stories and local knowledge about the region over the past three years. Our conversations provided vital understanding into the remarkable people and cultures of this area. Special thanks go to the following: Bernadette Yaayuk Alvanna-Stimpfle, Francis Alvanna, Wilfred Anowlic, Richard Beneville, Carol Gales, Ronald Engstrom, Howard Farley Sr., Neal Foster, Diana Haecker, John Handeland, Cussy Kauer, Joseph Kunnuk Sr., Marguerite LaRiviere, Charlie Lean, Dave McDowell, Nancy McGuire, Jack Omelak, Ray Paniataaq, Vince Pikonganna, John and Grace Pullock, Kitty Scott, and Lew Tobin.

Unless otherwise noted, all images are from the collection of the Carrie M. McLain Memorial Museum in Nome, Alaska. Bering Strait Inupiaq names for plants, animals, and objects appear in italics. It is hoped that this volume provides an essential resource to inform, engage, and inspire discussion on the historical legacy and cultural vitality of Nome and the Bering Strait region.

INTRODUCTION

The town of Nome stretches across the southwestern edge of the Seward Peninsula on the coast of Norton Sound and juts into the Bering Sea right below the Arctic Circle. The surrounding area features mountain slopes covered with wildflowers and berries, sheltered lagoons for birds and fish, and rolling tundra cut by lakes and rivers that feed into the icy sea. Alaska Native peoples have lived in the Bering Strait region for 4,000 to 6,000 years and continue to follow deeply rooted cultural beliefs and subsistence practices. The region is home to 20 villages and three culturally distinct groups of people, including Inupiat, Central Yup'ik, and St. Lawrence Island Yupik.

Hundreds of house depressions at the villages of Ayasayuk and Nuk around the Cape Nome area indicate substantial settlement of Alaska Native families along the southern coast of the Seward Peninsula for several hundred years. Construction of navigational improvements to the Nome harbor in 2005–2006 uncovered two partial sod houses and a midden at the mouth of the Snake River that also indicate Alaska Natives lived year-round at the current townsite of Nome before direct contact with Euro-Americans. During the late 1800s, Bering Strait families traveled along the coast to set up summer fish camps, attend trade fairs, and exchange Native-made items for Western goods from whalers and explorers.

The discovery of placer gold along tributaries of the Snake River in 1898 brought a stampede of miners, businessmen, and speculators to the Nome area. By 1900, steamships crowded the roadstead, and over 20,000 fortune-seekers were actively mining for gold, running well-stocked stores, serving up hot meals, and handling endless litigation. Thousands of prospectors fanned out over the Seward Peninsula to form mining camps at Council, Solomon, and Candle. The strikes resulted in enough gold to develop outlying settlements into unique towns with general stores, saloons, hotels, banks, and small hospitals. Bering Strait communities also participated in the gold frenzy on the Seward Peninsula. Many families traveled to Nome and camped under skin boats on the Snake River Sandspit where they carved ivory, made mukluks (boots) to sell, and worked as longshoremen for lightering companies.

Following the gold rush of the early 1900s, Nome experienced citywide fires, the collapse of the gold-mining industry, the 1918 influenza epidemic, and countless Bering Sea storms. Despite these hardships, the unique landscape and lifestyle of the Nome area has drawn together the community of 3,700 residents. Long, sun-drenched summer days encourage outdoor excursions to pick wildflowers and berries, watch for migratory birds, fish the rivers, and comb the sandy beaches for sea glass. The arrival of frosty winter brings opportunities to cross-country ski, watch a local snowmachine race, see the Northern Lights, and cheer on mushers as they cross the finish line of the Iditarod Trail Sled Dog Race. A commitment to the community from residents, the central position of the town for increased shipping and scientific exploration in the Arctic, and continued development of rich natural resources all point to the persistence of Nome as an active crossroads on the Bering Sea.

One

Seasonal Cycle

The Bering Strait region provides a rich habitat for marine and land mammals, birds, and fish. Communities have maintained a close relationship with the changing environment here for generations. By following a seasonal cycle, residents make efficient use of subsistence resources at different times of the year.

Winter covers Nome and the Bering Strait in a blanket of white with snowfall averaging around 76 inches a year. The amber sun hangs low on the horizon, and the Northern Lights dance across nighttime skies. In clear weather, hunters set off after wolverine, wolf, and fox despite temperatures hovering between 5°F and 10°F. Fishermen catch tomcod and crab through holes in shore-fast sea ice and pike and trout through river ice.

Spring brings increased daylight hours and warms the tundra, encouraging the growth of nutritious willow leaves, beach greens, and tubers from potato plants. By late April, spring breakup is well underway, and open leads through the ice allow boats to pursue seals, walrus, beluga, and bowhead whales as they migrate through the Bering Strait to reach summer feeding grounds in the Arctic Ocean. Migrant seabirds and waterfowl arrive, and hunters gather eggs to be boiled or stored in seal oil.

Summer skies resonate with the calls of over 160 species of migratory birds as they raise their young and feast on plants and insects among the protected lagoons and tundra ponds. The summer solstice in June provides 21.5 hours of daylight followed by long sunny days of berry picking, gathering sourdock, and fishing for grayling, whitefish, and salmon. Summer travel by skin boats to coastal fish camps provided families in the past with critical food sources as well as opportunities to trade and catch up with neighbors.

Autumn paints glorious colors across the tundra meadows and lush riverbanks. In September, flocks of cranes and geese embark on their southward migration, and ptarmigan begin to show their white plumage. Hunters trek north across the Seward Peninsula in search of the Western Arctic caribou herd, and Native herders harvest domesticated reindeer. During late autumn, brown bears roam across the rolling hills in their voracious search for food to pack on a thick layer of fat for winter hibernation. Expansion of the Arctic Ocean ice cap forces marine mammals to return south through the Bering Strait where they provide coastal communities with food and materials for the coming winter.

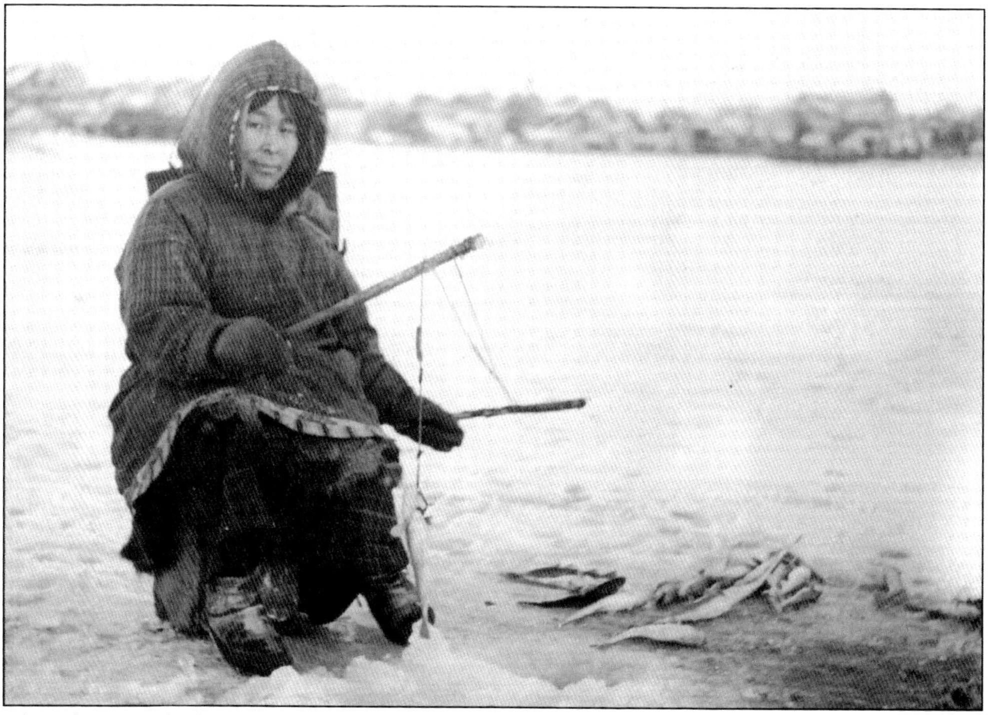

This photograph shows an Inupiat woman jigging for tomcod (*iġałuaq*) through a hole in the shore-fast ice outside of Nome around 1905. Tomcod escape predatory seals by entering lagoons and seeking shelter under the ice. The lagoon at Safety Sound is a perfect spot for Nome residents to chip holes in the ice and jig for tomcod that can be enjoyed frozen, dried, or baked. (Photograph by B.B. Dobbs; Carrie M. McLain Memorial Museum, 1992.2.232.)

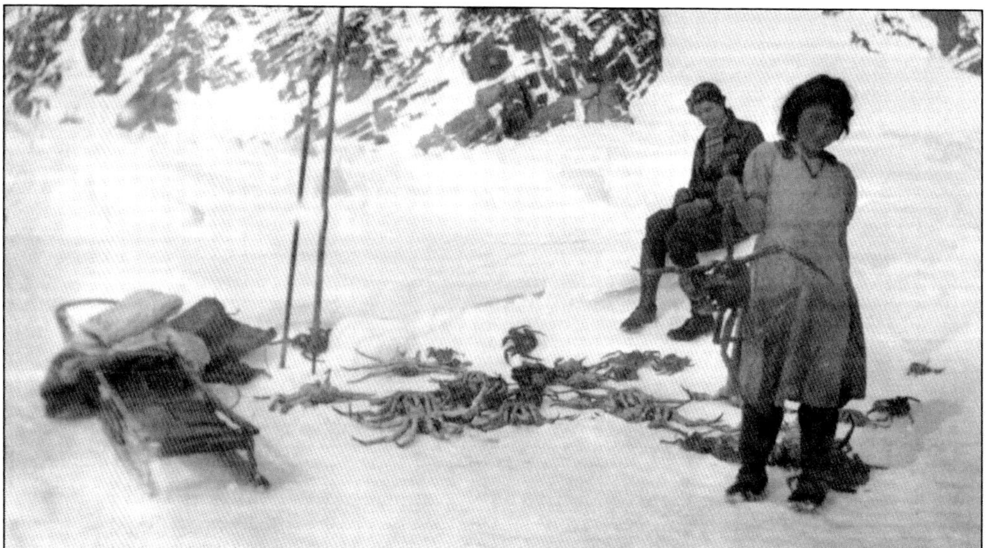

A school teacher and young girl catch king crab (*qaqquq*) on hand lines through the ice surrounding King Island around 1930. Standing upright in the snow are an ice tester and a netted scoop used to keep fishing holes clear of slush. The small pack sled and sealskin bag will be used to transport the day's catch back to the village. (Carrie M. McLain Memorial Museum, McLain-603.)

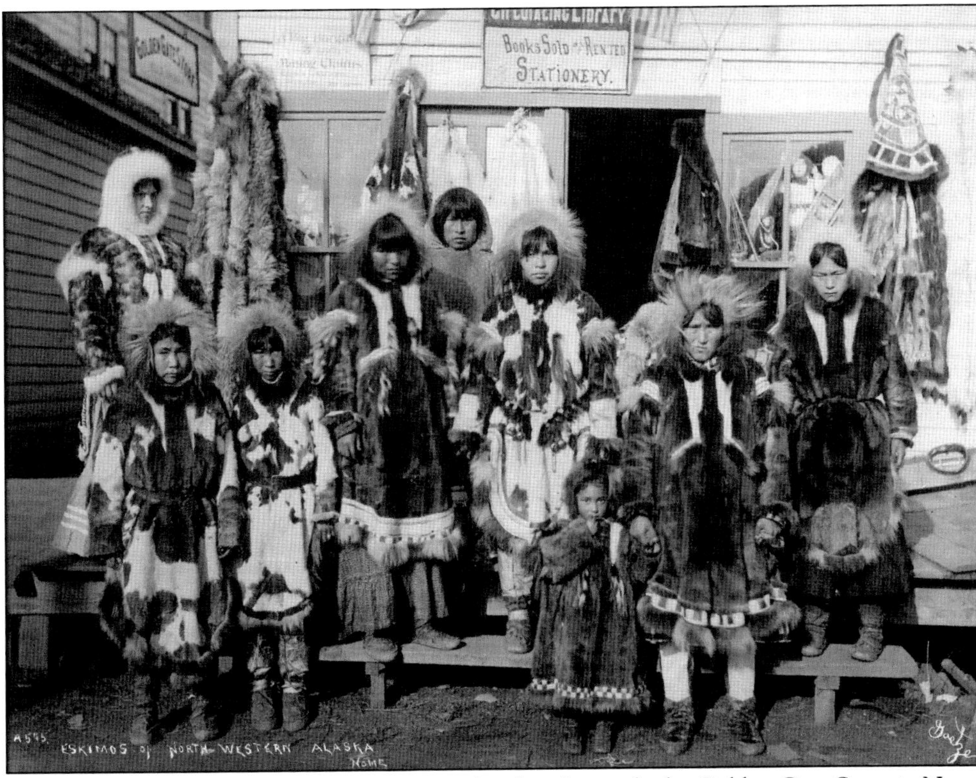

Women and children pose in fancy winter parkas (*atigi*) outside the Golden Gate Store in Nome around 1905. Bering Strait winter clothing features carefully arranged designs of skins from caribou, fox, wolf, wolverine, ground squirrel, seal, lynx, and Arctic hare. Skins are soaked and scraped until pliable enough to be measured and cut with a small sewing ulu (*kimaliq*); they are then sewed together using bodkins, needles, and thread, which is traditionally made from caribou sinew (*ivalu*). (Photograph by Otto D. Goetze; Carrie M. McLain Memorial Museum, 1979.1.234.)

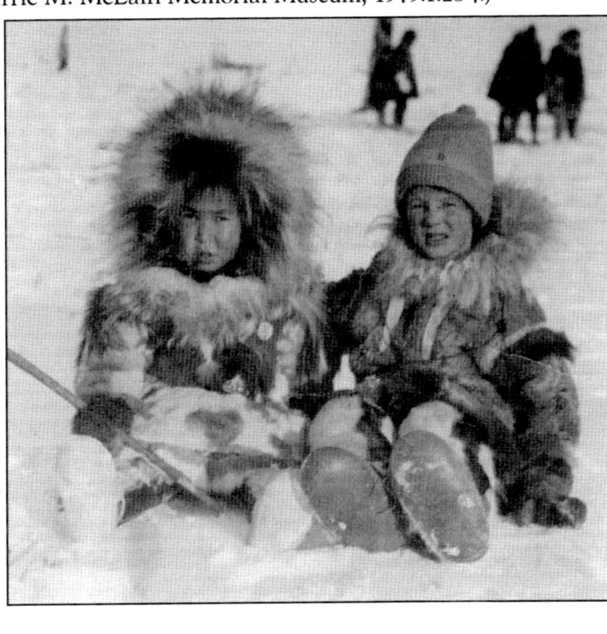

Sara Shields (right) and a friend bundle up in warm fur parkas and mukluks for a tomcod-fishing excursion out on the ice around 1907–1908. Many Inupiat women in Nome made fur parkas and mukluks to sell to non-natives who valued the exceptional warmth and unique beauty of the hand-sewn clothing. The Nome Skin Sewers Cooperative Organization was organized by sewers in 1940, and for over 20 years, its members served to manufacture, buy, and sell fur clothing across Alaska. (Photograph by Walter Shields; Carrie M. McLain Memorial Museum, 2017.5.)

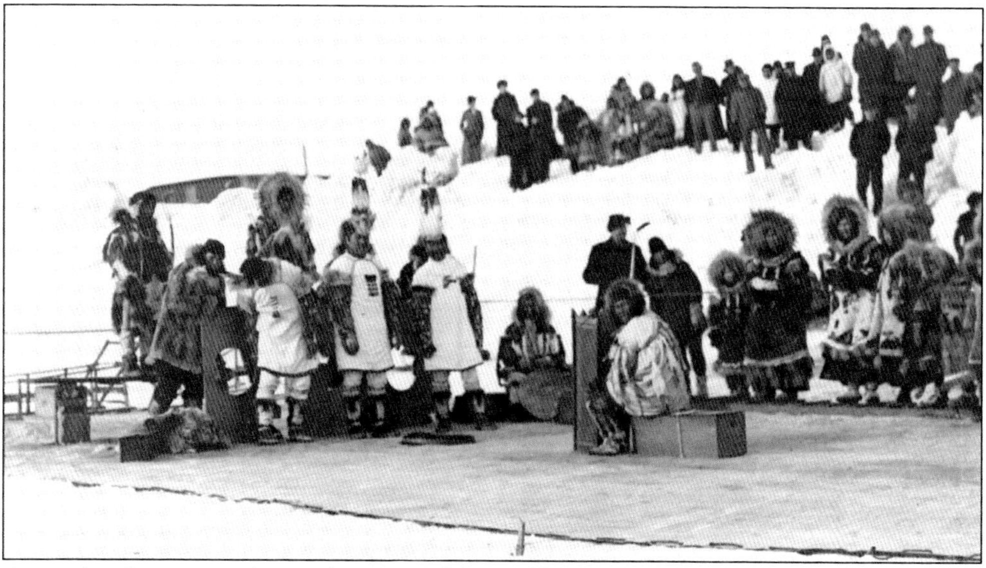

A crowd gathers to watch a performance of the Wolf Dance on the ice outside of Nome around 1910. The Wolf Dance recounts the story of Eagle Mother, who teaches a man how to hunt, dance, and sing, and the correct method to prepare a Messenger Feast (*Kivġiq*) in order to host guests. (Carrie M. McLain Memorial Museum, Karrer-39.)

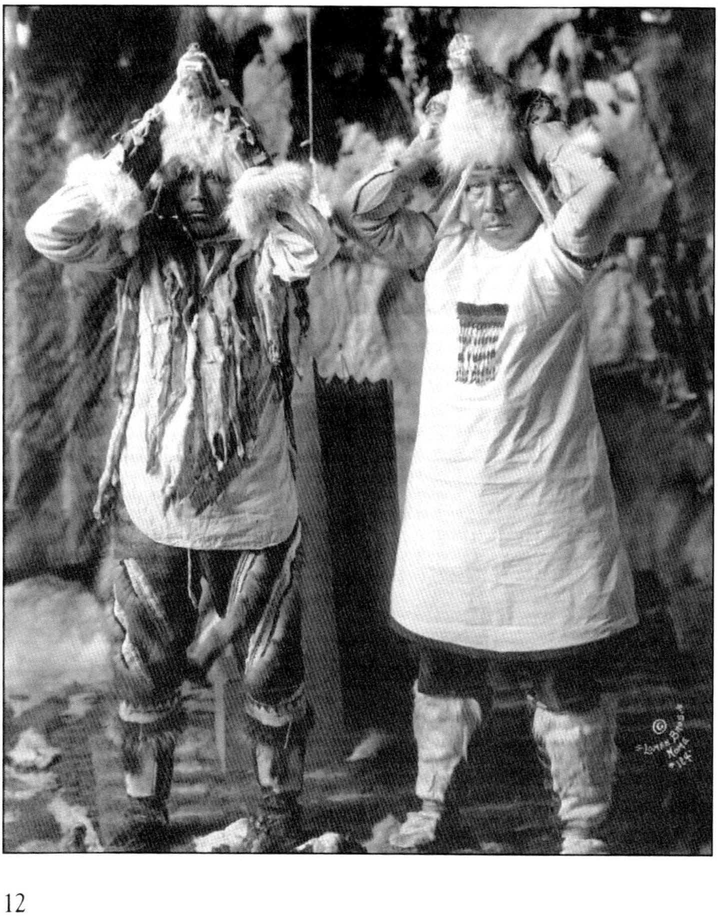

Inupiat dancers from Wales pose in Wolf Dance regalia inside the Lomen Brothers studio in Nome around 1910. The complex performance features special eagle-feather headdresses (*suluktat*), wolf skins, rattle mittens (*aqłitiik*), feather wands, and a box drum (*kalukaq*). (Carrie M. McLain Memorial Museum, 2005.1.)

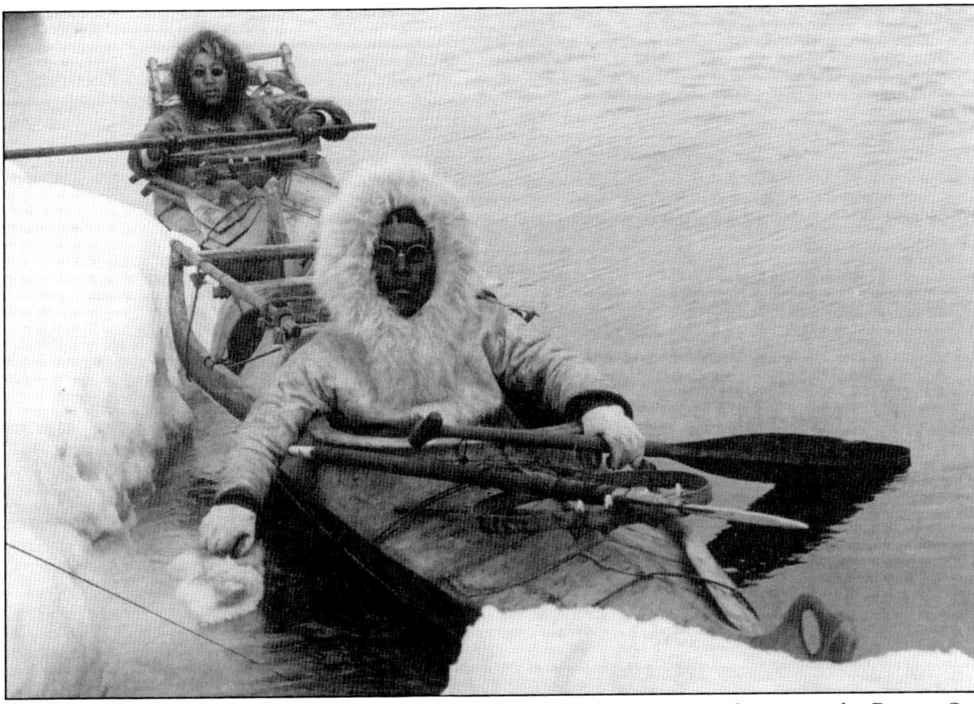

Seal hunters from King Island are well prepared for variable spring conditions in the Bering Sea around 1910. Sealskin-covered kayaks, wood paddles, pack sleds, ice testers, harpoons, circular frames to hold the harpoon line, fur parkas, and ice goggles make up their specialized hunting kits. (Photograph by Lomen Brothers; Carrie M. McLain Memorial Museum, 2005.1.745.)

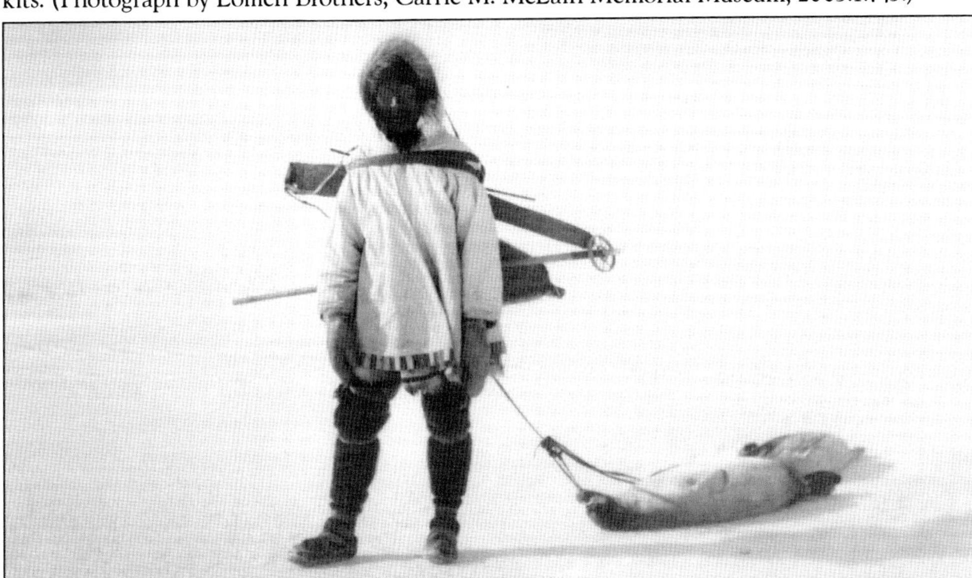

An Inupiat hunter drags two seals over the ice using a line attached to a strap around his chest outside of Nome around 1910. A sealskin hunting bag, ice tester, and rifle case are slung across his back. Hunters traditionally wore goggles made of ivory or bone to protect their eyes from the harsh glare of the ice, and sealskin pants and boots to keep their legs and feet warm. (Carrie M. McLain Memorial Museum, Bell-N-48.)

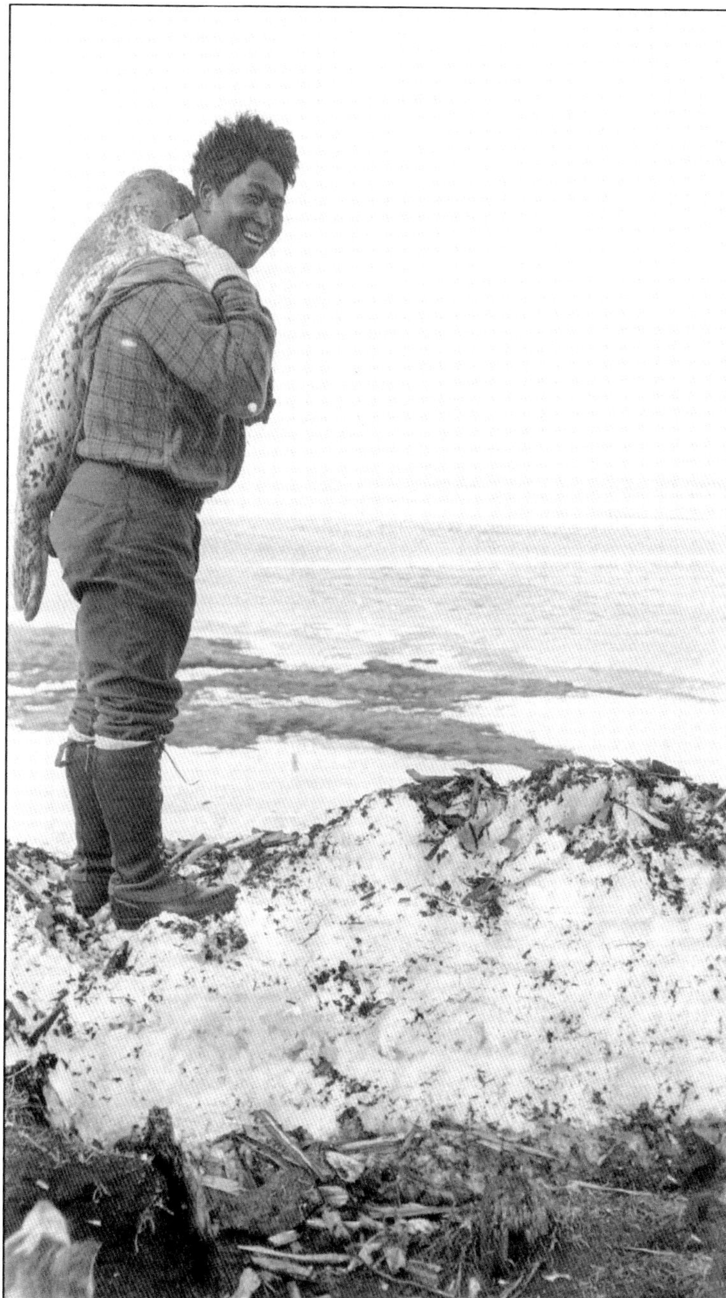

A hunter carries a spotted seal (*niqsaq*) across the spring tundra around 1930. Spotted seals feed on herring, capelin, and cod and will follow these fish into the ice-free waters of Norton Sound. Bering Strait coastal communities hunt seals from the edge of sea ice or holes during the winter. By April, hunters follow open leads in the ice to pursue seals on their northward migration through the Bering Strait. Seals were traditionally taken with harpoons and retrieved with a wooden float with hooks (*manaq*). Seal meat can be frozen or dried and was traditionally placed with cubes of blubber into sealskin bags or "pokes" that kept the meat and oil edible for a year. (Carrie M. McLain Memorial Museum, McLain-710.)

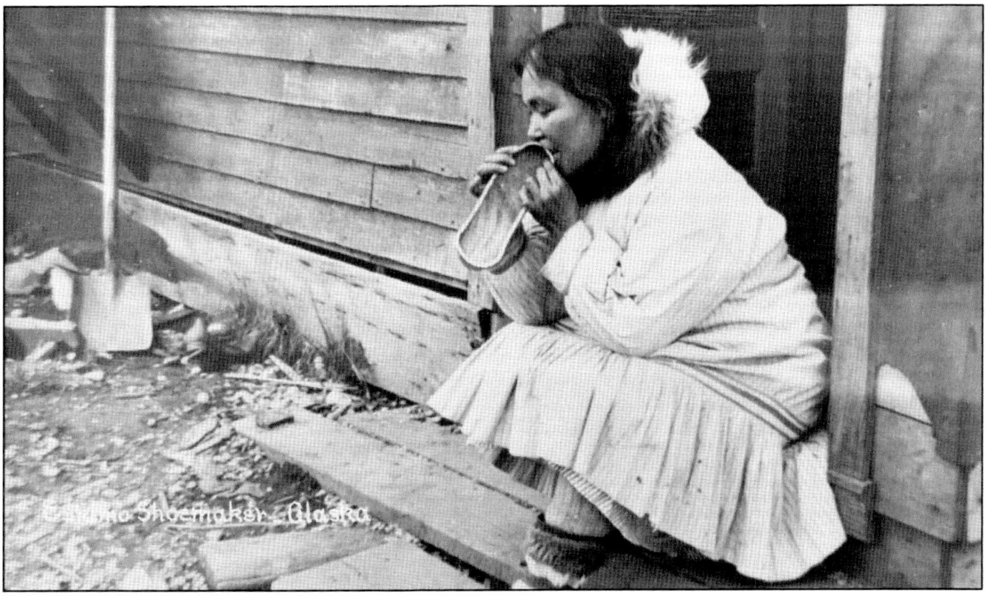

This photograph, taken in Nome between 1941 and 1946, shows an Inupiat woman using her teeth to crimp bearded sealskin into a hard sole, which will be used for a new pair of mukluks. Sewers removed, cut, and tanned sealskins in specific ways depending on their intended use as rope, boat covers, floats, parkas, pants, or footgear. (Photograph by Elva Bedford; Carrie M. McLain Memorial Museum, 2002.17.50.)

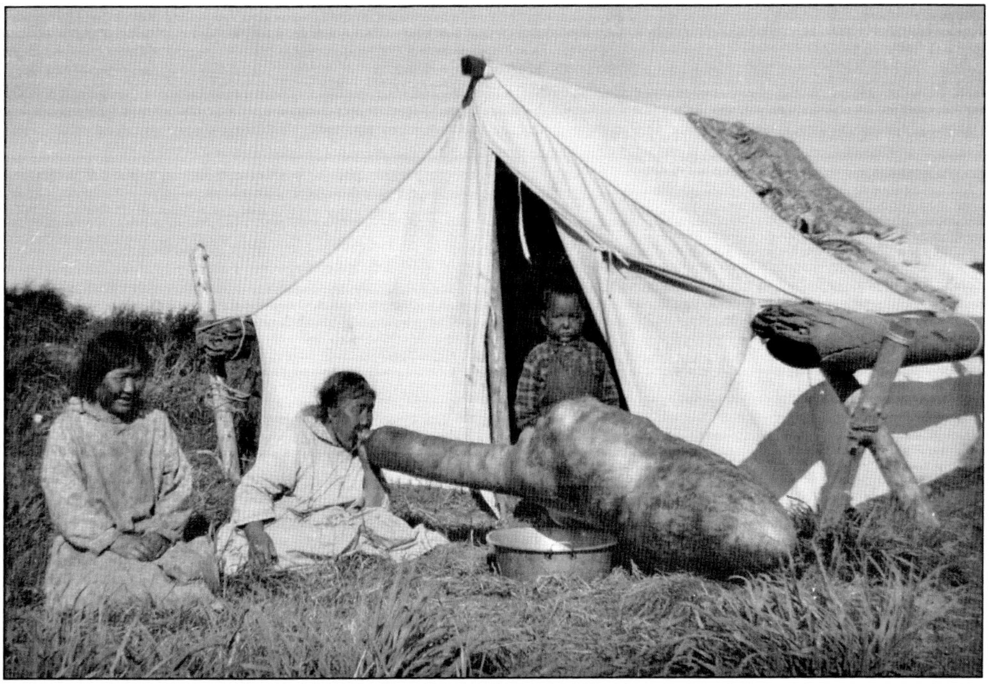

An Inupiat woman inflates a sealskin with the hair removed at a family's summer camp on the Seward Peninsula between 1941 and 1946. The sealskin will be turned into a float and used when hunting walrus or whales in a skin boat (*umiaq*). (Photograph by Elva Bedford; Carrie M. McLain Memorial Museum, 2002.17.15.)

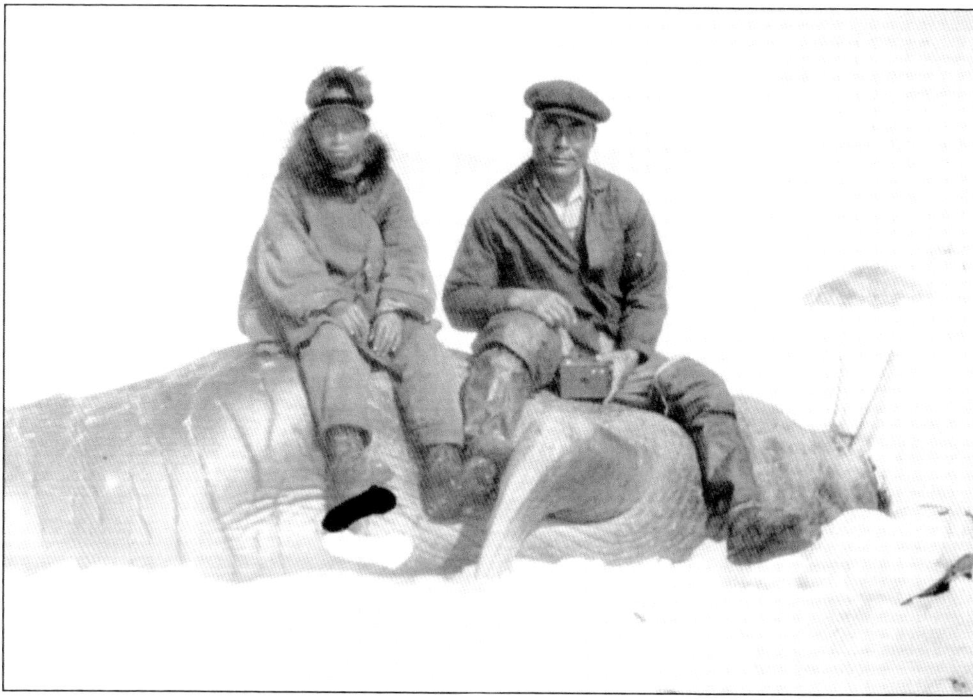

Two hunters sit on top of a walrus (*aiviq*) out on the Bering Sea ice around 1930. During spring breakup, Alaska Native hunters pursue walrus resting on ice cakes, swimming in the water, or hauled out on the beach. Pacific walrus average around 2,500 pounds with two prominent ivory tusks that can grow up to 36 inches in length. (Carrie M. McLain Memorial Museum, McLain-703.)

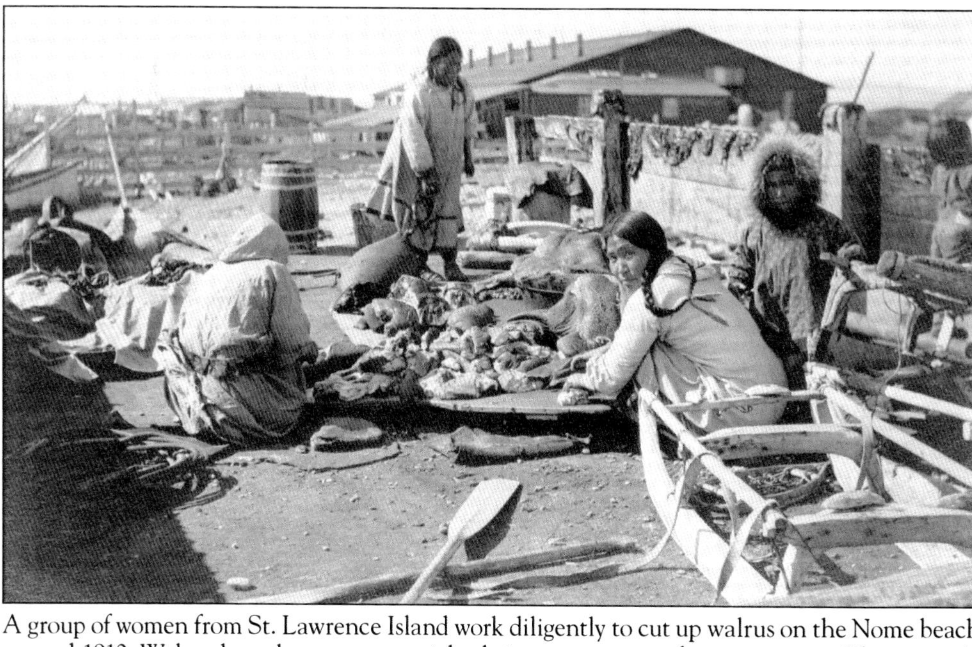

A group of women from St. Lawrence Island work diligently to cut up walrus on the Nome beach around 1910. Walrus have been an essential subsistence resource for generations. They provide meat for people and dogs, skins for boat covers, intestines for waterproof garments, and ivory for tools and artwork. (Carrie M. McLain Memorial Museum, 2017.5.)

A young hunter proudly shows off his catch of four ducks during a late spring day on the Seward Peninsula around 1930. Ducks, geese, and cranes were traditionally hunted with bolas or arrows tipped with ivory or antler. Migratory birds arrive in the Nome area every May and June and join about two dozen resident species that tough out the long Arctic winters every year. Three roadways leading out of Nome provide exceptional access to a colorful range of songbirds, shorebirds, and seabirds for birdwatching, research, and subsistence hunting. (Carrie M. McLain Memorial Museum, McLain-668.)

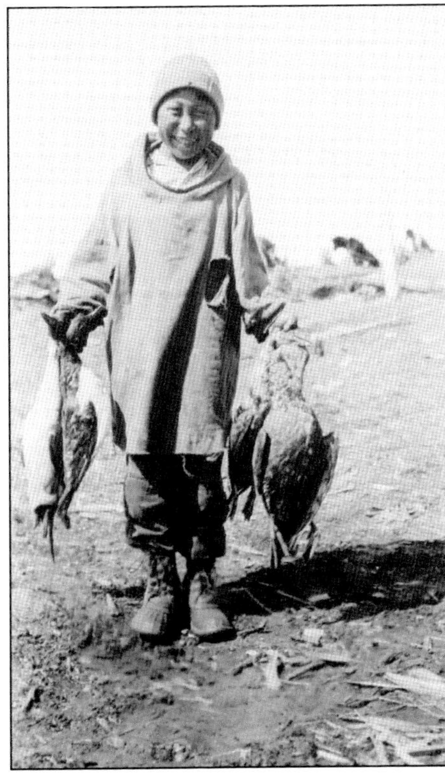

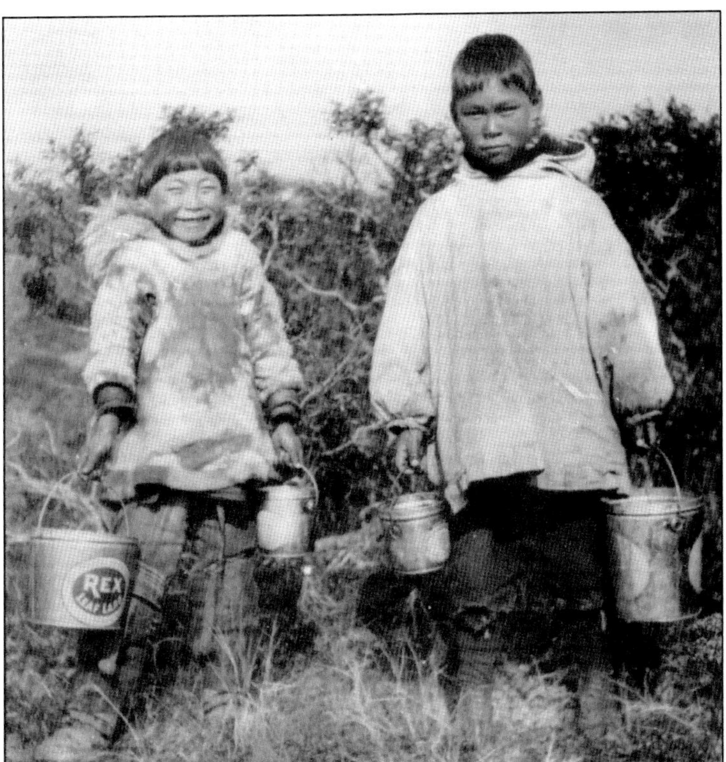

Two young boys get ready to head across the tundra to fill their Rex lard pails with warm summer berries around 1905. Fresh berries could be packed with seal or fish oil into sealskin pokes that kept the berries edible through the following spring. (Carrie M. McLain Memorial Museum, 1979.1.528.)

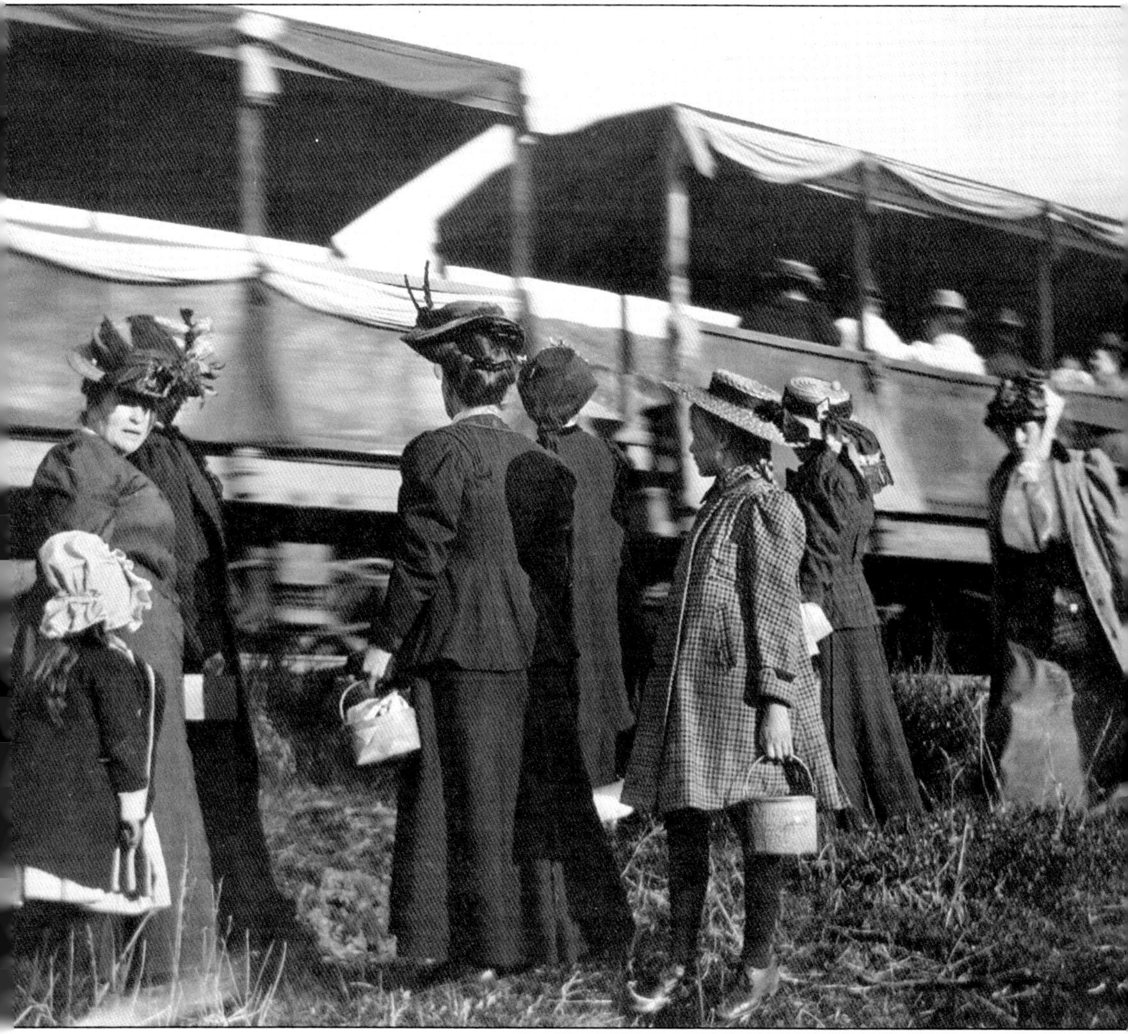

Clara Mielke (in white bonnet at left) holds the hand of her mother, Lizzie, as they wait to board the Seward Peninsula Railroad for a berry-picking excursion on the tundra around 1905. Picking berries continues to be a favorite family activity in Nome. Local knowledge about areas with abundant salmonberry (*aqpik*), blueberry (*suga*), crowberry (*aluk*), and cranberry (*kikmik*) plants is shared among family members and close friends. Residents often enjoy berries in a dessert called blueberry delight, mixed together and served with pilot bread and tea, or combined with whipped caribou tallow, sugar, and seal oil to make *agutaq*, or "Eskimo ice cream." (Carrie M. McLain Memorial Museum, 1979.1.251.)

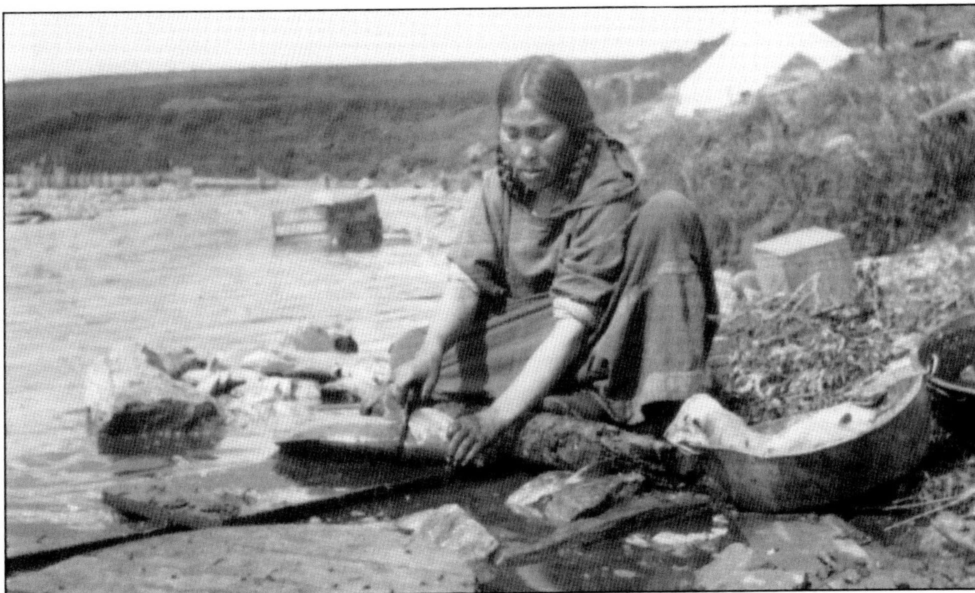

An Inupiat woman cleans salmon (*iġalugruaq*) with an ulu along the river at a summer fish camp between 1903 and 1909. Racks set up next to the tent hold rows of cut and drying salmon. Subsistence fishermen in the Nome area typically catch salmon with a rod and reel, a seine net pulled through the water, or a gill net anchored with sinkers. (Photograph by Arthur L. Bell; Carrie M. McLain Memorial Museum, Bell-N-61.)

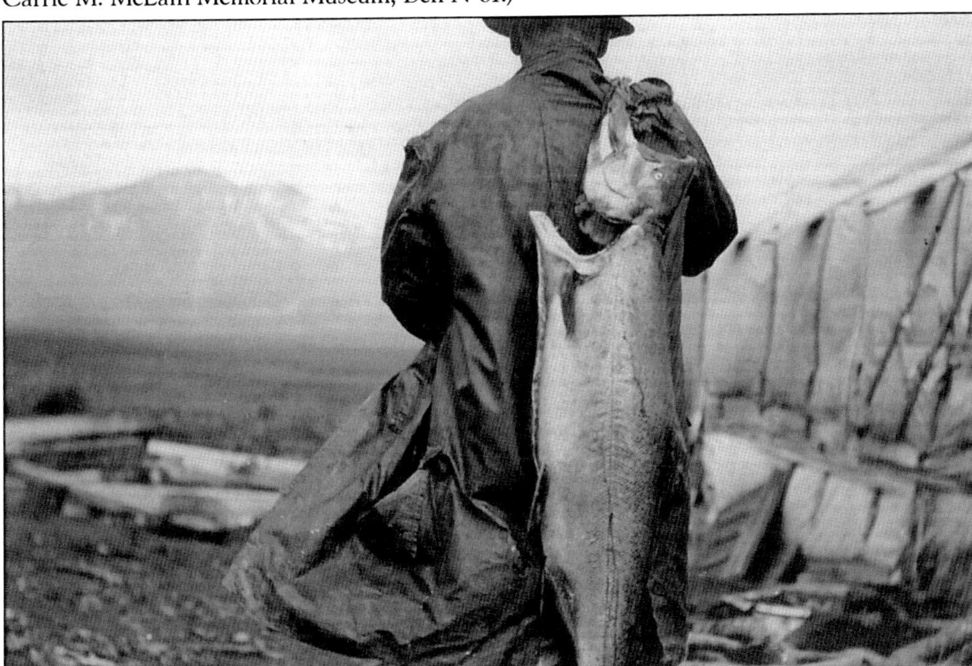

A successful fisherman carries home a Chinook salmon caught in Iron Creek, a southern tributary of the Pilgrim River that originates at Salmon Lake, located 40 miles north of Nome, around 1910. The summer waterways of the Seward Peninsula teem with all five species of Pacific salmon: chum, chinook, pink, sockeye, and coho. (Carrie M. McLain Memorial Museum, Rolando-118.)

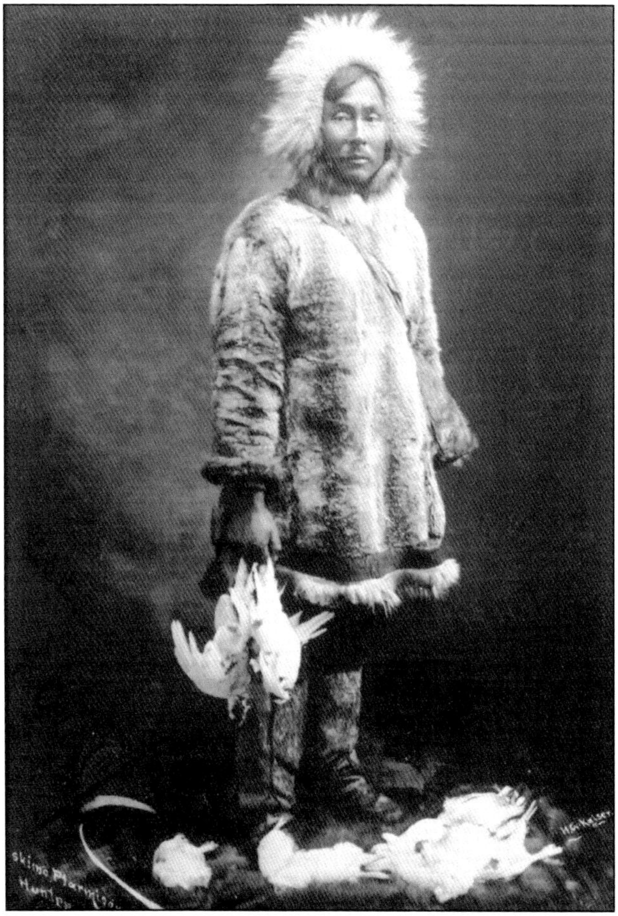

Surrounded by ptarmigan, a hunter poses for a studio portrait in Nome wearing his ground squirrel parka, gun case, and sealskin boots. Ptarmigan were traditionally hunted with bows and arrows with blunt tips to kill the bird without damaging the meat. (Photograph by H.G. Kaiser; Carrie M. McLain Memorial Museum, NMP-20-665.)

Young hunters pose with their rifles and catch of ptarmigan and Arctic hare on the fall tundra outside Nome around 1915. The stout and social willow ptarmigan (*aġagriq*) changes from light brown in summer to snowy white in winter to protect itself from predators. Ptarmigan are hunted for their dark meat, which tastes savory roasted and in stews. (Carrie M. McLain Memorial Museum, 1979.1.350.)

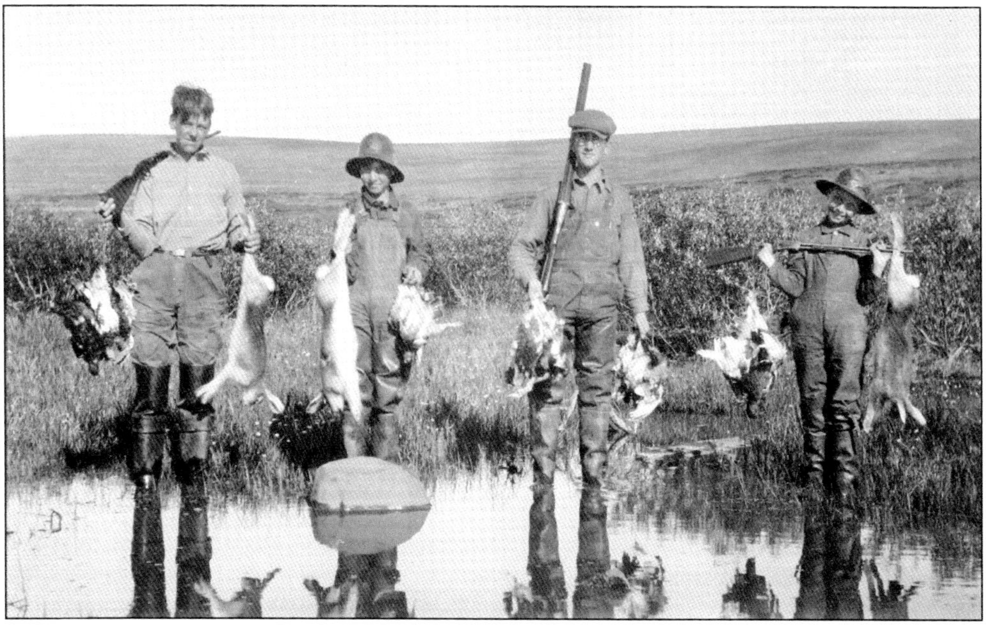

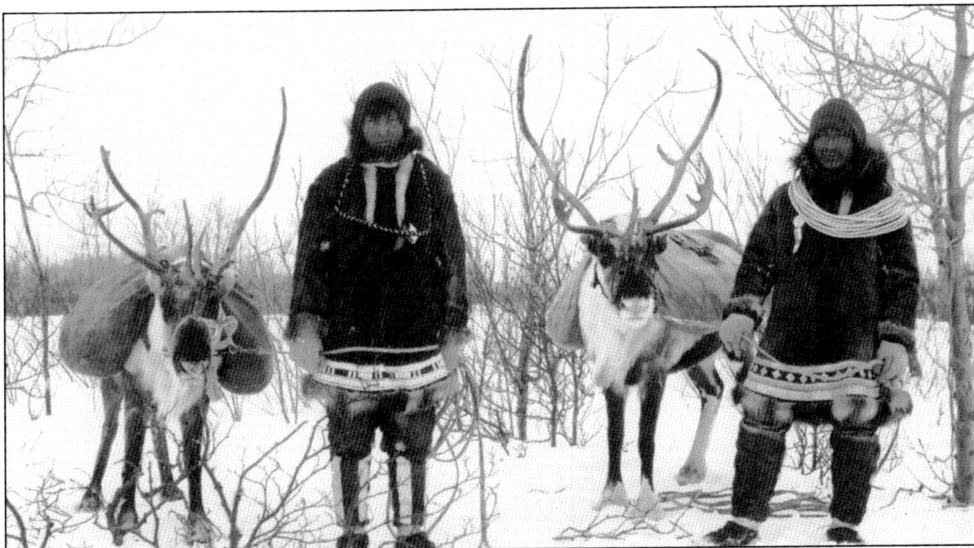

Inupiat herders lead their pack reindeer through the snow and willows on the Seward Peninsula around 1915. Reindeer feed on grasses, willow buds, and berries during the summer and subsist almost entirely on lichen (reindeer moss) during the winter. (Carrie M. McLain Memorial Museum, 1992.2.215.)

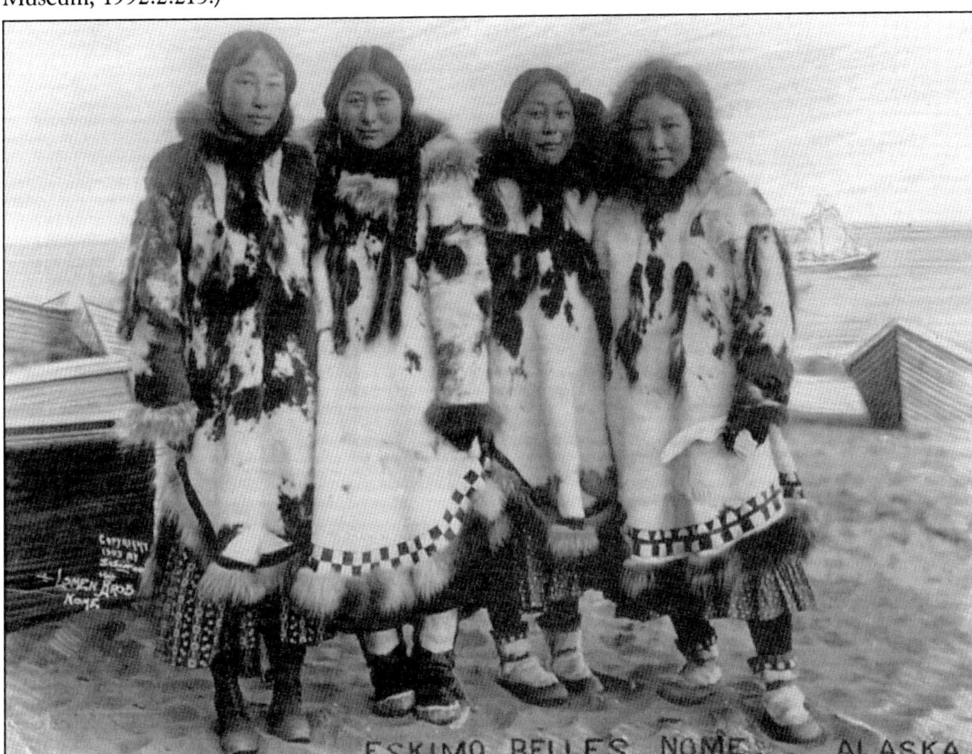

Young women from Wales pose on the sand in front a backdrop painted with a scene from the Nome beach in 1903. Spotted skins of reindeer provide natural decoration for parkas while hair from the legs makes good material for mukluks. (Photograph by Lomen Brothers; Carrie M. McLain Memorial Museum, Rolando-34.)

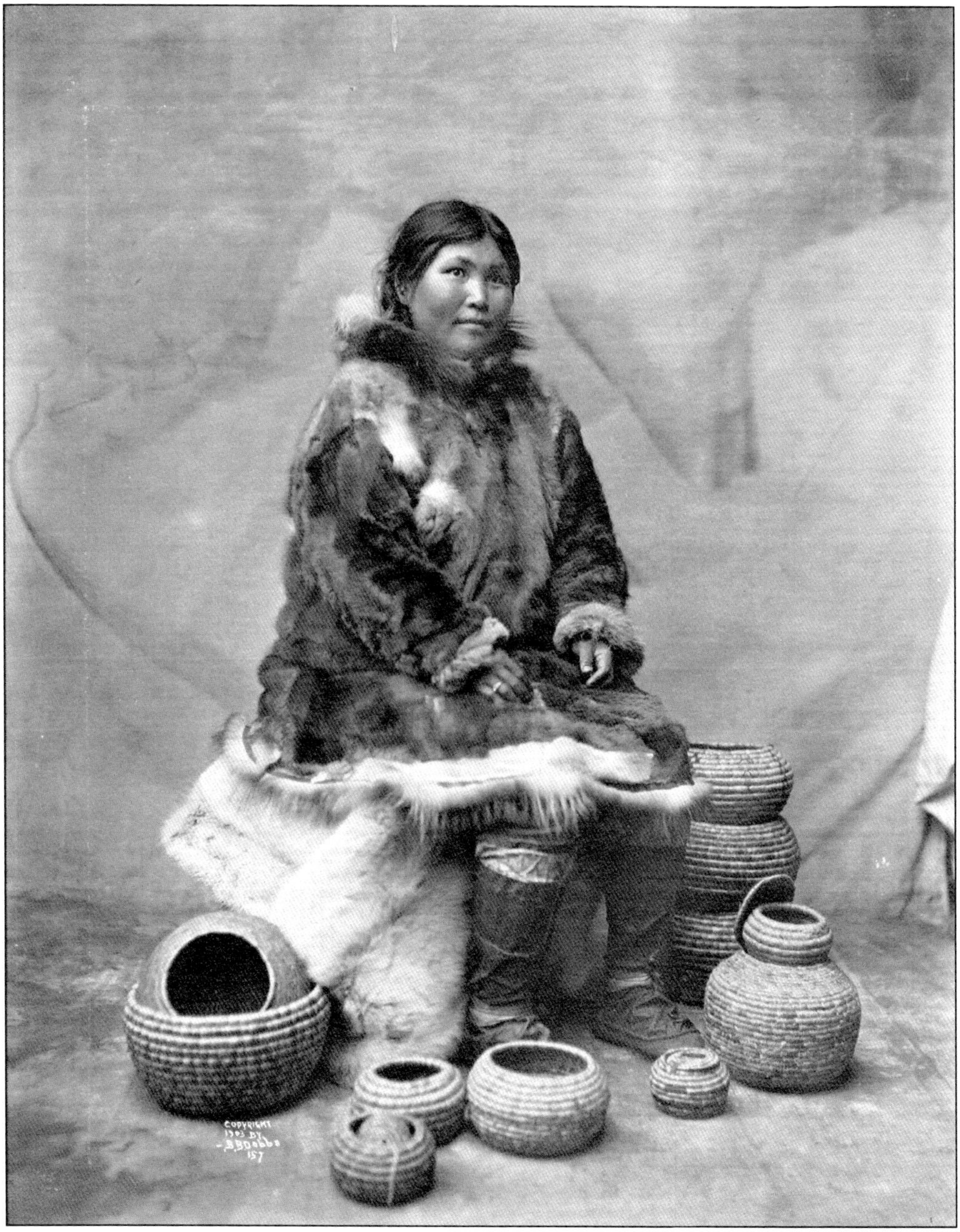

An Inupiat woman poses in a fur parka and waterproof boots surrounded by coiled grass baskets in Nome in 1903. Weavers typically gather basket grass (*ivgich*) in the fall after the first frost. The grass is then soaked, split, and woven. Grass baskets with beads, feathers, or ivory attachments became popular tourist items in Nome during the early 1900s. The coiled basketry tradition was almost completely abandoned in the Bering Strait area after the 1918 influenza epidemic. Coiled baskets today derive mainly from southwest Alaska, where Yup'ik artisans weave in creative designs and form coils into sculptures of dolls and animals. (Photograph by B.B. Dobbs; Carrie M. McLain Memorial Museum, 2013.1.27.)

Two

Tent City

In 1897, Daniel Libby mounted an expedition to the Seward Peninsula and discovered gold the following spring at Mesling and Ophir Creeks, which led to the founding of Council City. That summer, Jafet Lindeberg, Erik Lindblom, and John Brynteson set out from Council on an old flat-bottomed boat and traveled west to the mouth of the Snake River a few miles from the modern site of Nome. The three prospectors struck gold on the rich Anvil Creek and staked claims on five tributaries of the Snake River.

Discovery of placer gold in 1898 by the "three lucky Swedes" brought a stampede of miners who arrived that winter by dogsled from the exploited gold fields of Dawson and those who crowded onto steamships from Seattle and San Francisco during the following spring of 1899. Newcomers camped on the beach and, in July 1899, discovered rich deposits of gold in the sand beneath their feet. News of the golden sands spread quickly, and in 1900, the population swelled from 12,000 in the spring to more than 20,000 by midsummer.

Early arrivals discovered that they needed a sturdy canvas wall tent to store supplies and provide essential protection from the rain and snow. Tents helped secure mining claims for the first prospectors and were carefully guarded so another miner did not "jump his claim" by pulling up tent stakes and putting in his own. Bering Strait families from King Island, Little Diomede, Wales, and Gambell and Savoonga on St. Lawrence Island, arrived to the active shore and set up summer camps with skin boats and tents on a mile-long strip of land known as the Sandspit at the mouth of the Snake River.

The tent city of Nome grew to five miles along the coast with only a narrow wagon path to separate the tightly packed rows of multipurpose structures. General stores, restaurants, bathhouses, and other facilities were all housed in canvas. Smoke from campfires, the strong scent of horses, and the aroma of bread from the California Bakery wagon drifted among the tents and mixed with sounds from the busy shorefront.

The rapidly expanding settlement of tents would be known as Anvil City, after a nearby mountain with a large, anvil-shaped rock on its summit, until the fall of 1899, when a consent government voted to change the name to Nome.

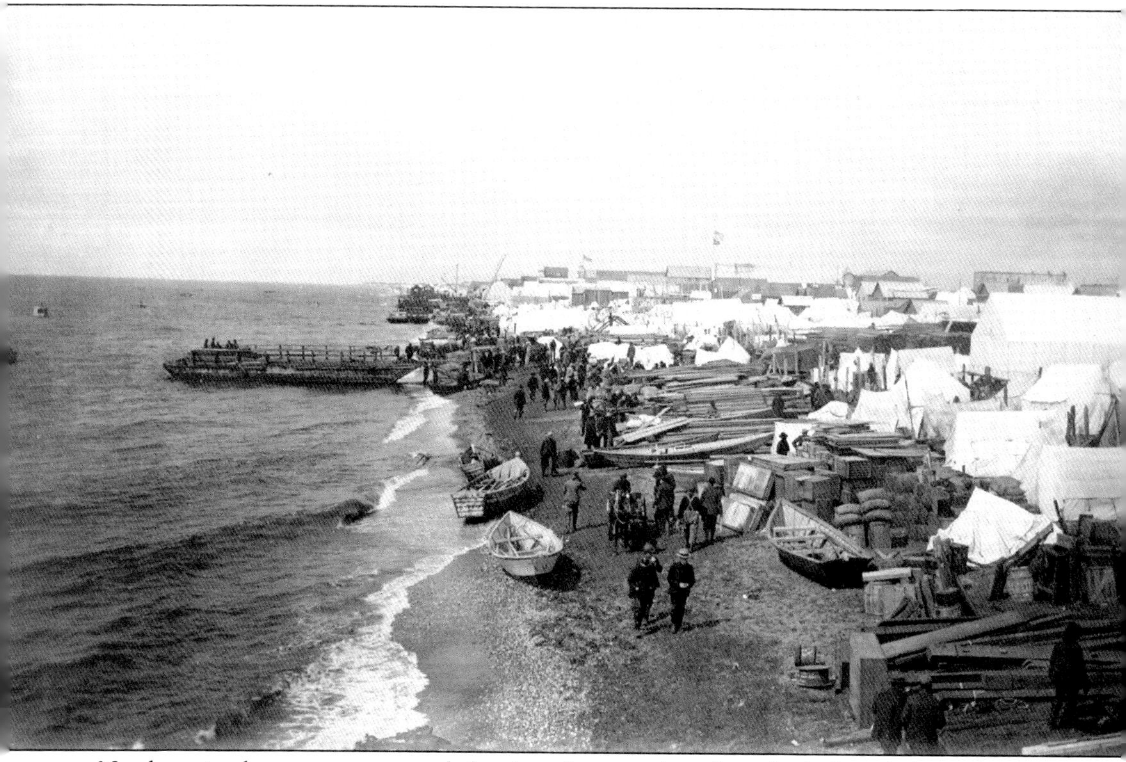

Newly arrived prospectors crowd the shorefront as they disembark from lighters, look over supplies, and prepare to set up tents during the summer of 1900. Just out of reach of the waves are precarious stacks of lumber, doors with plate-glass windows, and bales of straw. Small scows for transporting freight up and down the coast can be seen on the beach and in use on the water. More substantial structures along Front Street appear to the north with American and Alaska Commercial Company flags waving and a flapping banner advertising the *Gold Digger* newspaper. (Carrie M. McLain Memorial Museum, 2006.21.)

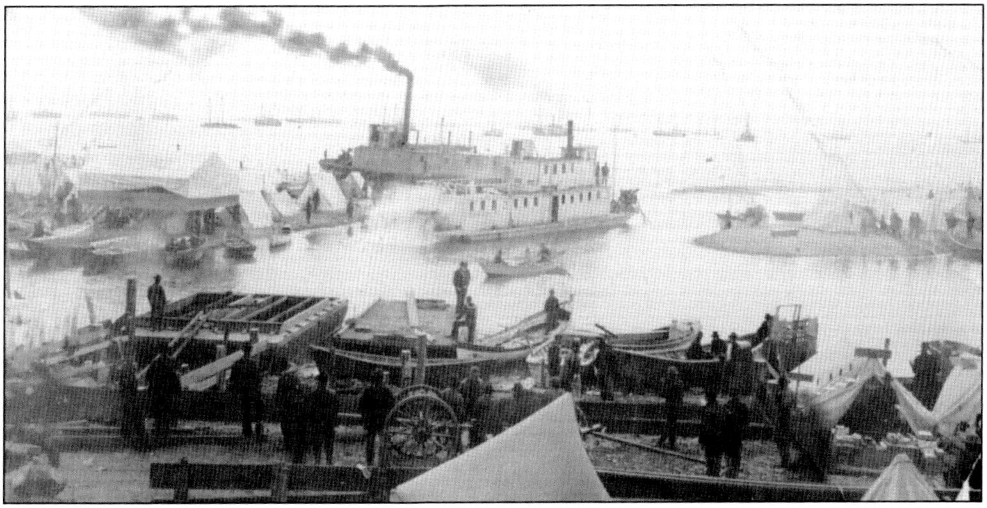

The mouth of the Snake River hums with activity in 1900 as miners visit on the Sandspit, a merchant peddles supplies from the front of his tent, and men gather to watch the mosquito fleet of small boats load cargo to trade along the coast. Two shallow-draft sternwheelers blast smoke as they prepare to head back out into the deep water of Norton Sound. Over 20 steamships carrying passengers and freight can be seen anchored in the roadstead. (Carrie M. McLain Memorial Museum, NMP-17-501.)

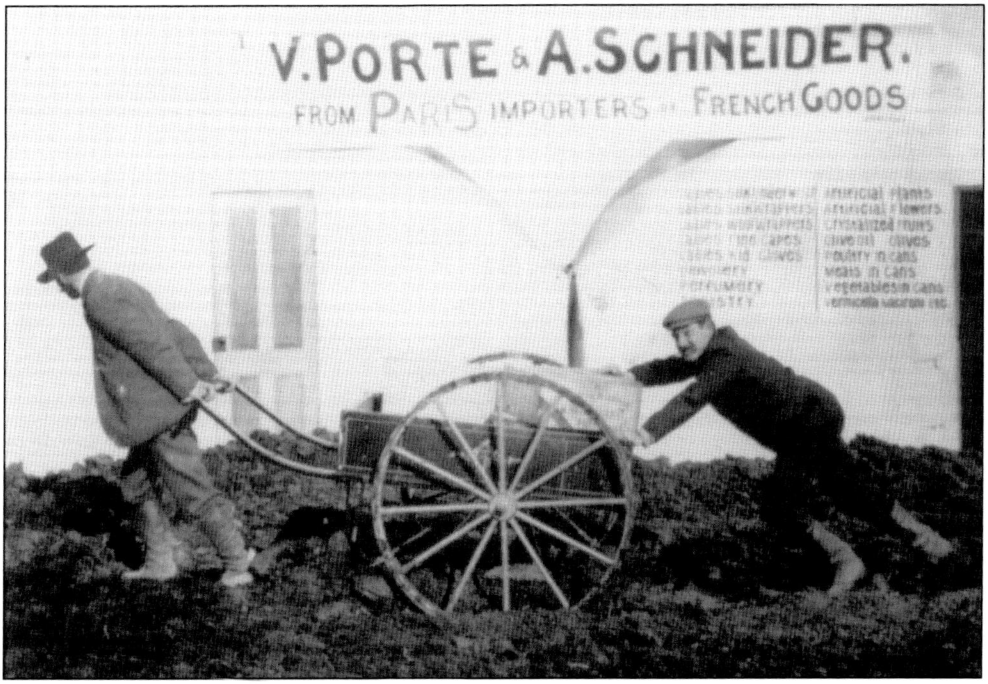

Storeowners V. Porte and A. Schneider push a load of merchandise through the deep mud of a Nome street in 1900. Their store catered to a growing number of fashionable women with a diverse offering of luxury goods, including ladies' silk underwear, capes and gloves, jewelry, perfume, artificial plants and flowers, and crystal fruits. (Carrie M. McLain Memorial Museum, NMP-96-5-43.)

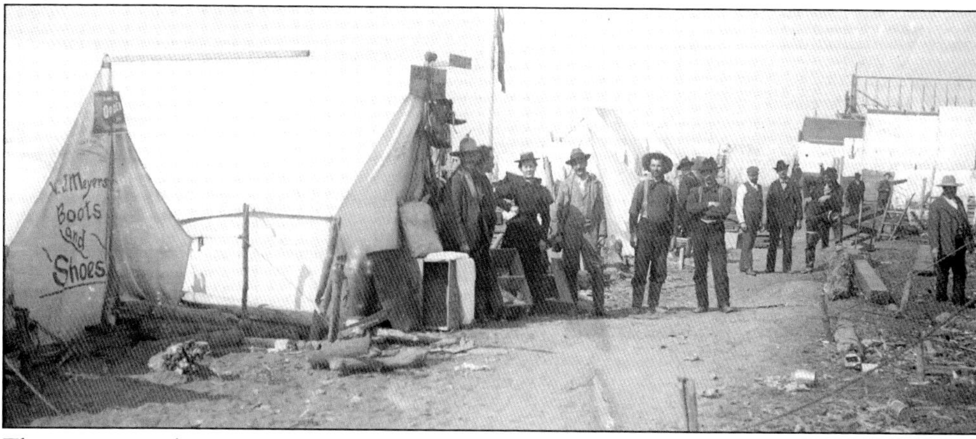

This street view shows a tent offering boots made to order and a grocer selling bread and vegetables. Note that narrow roads wound between tents pitched side by side and back to back. Nome was a sleepless town during the summer of 1900 as tension from the unfamiliar environment, crowded conditions, and haste to find gold ran through the tent city. (Carrie M. McLain Memorial Museum, McLain-56.)

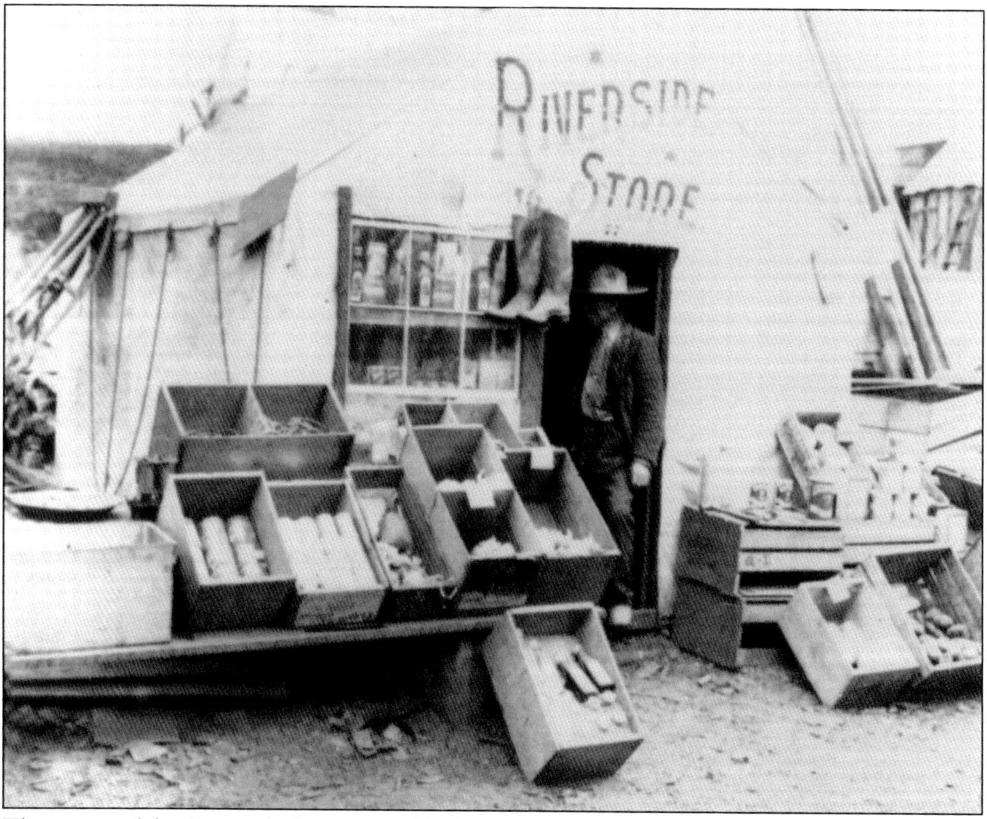

The owner of this Riverside Store wasted little time in pitching a tent and putting out displays of canned goods, crates of vegetables, pairs of boots, and gold pans in 1900. Successful tent merchants went on to build storefronts along the main thoroughfares of Nome, while those less fortunate packed up their belongings at the end of summer and boarded a steamship back home. (Carrie M. McLain Memorial Museum, NMP-96-5-31.)

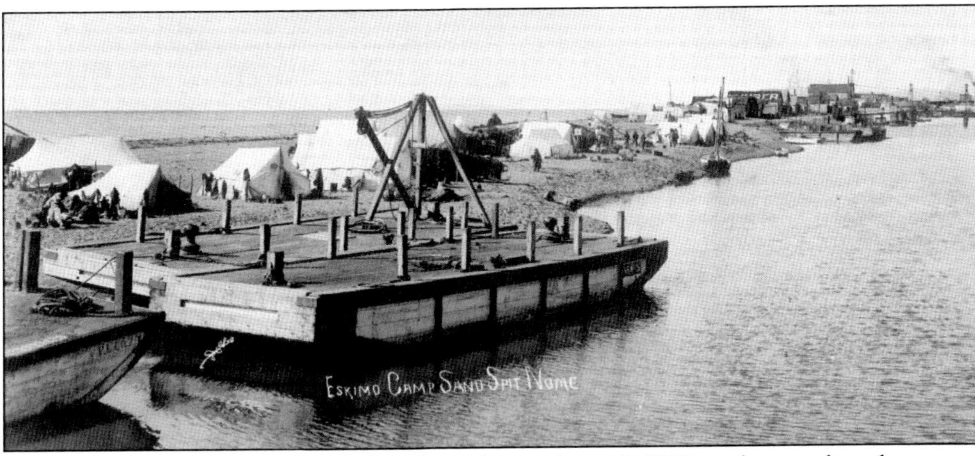

Many Bering Strait families traveled to Nome during the early 1900s and camped on the narrow Snake River Sandspit. Barges are shown here next to a busy scene on the Sandspit in 1905 with men carving ivory, clothes drying in the sunshine, and children playing at the water's edge. (Photograph by B.B. Dobbs; Carrie M. McLain Memorial Museum, 96.26.69.)

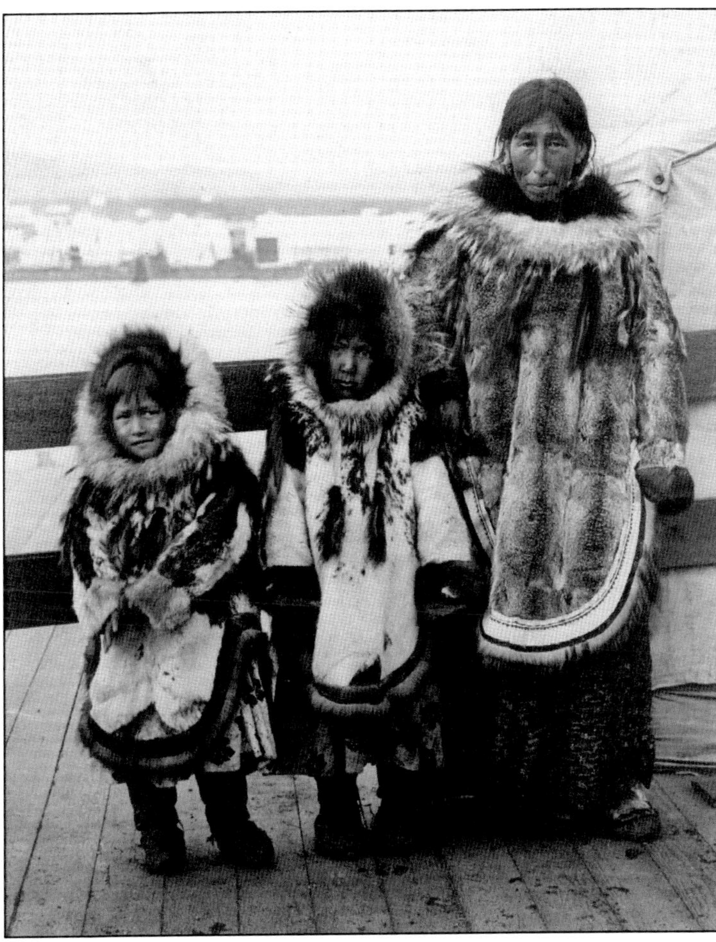

Wearing parkas made from ground squirrel and reindeer skins, an Inupiat woman and two young girls pause for a photograph on a footbridge leading to the Sandspit in 1900. Rows of white canvas tents can be seen in the background. (Carrie M. McLain Memorial Museum, McLain-541.)

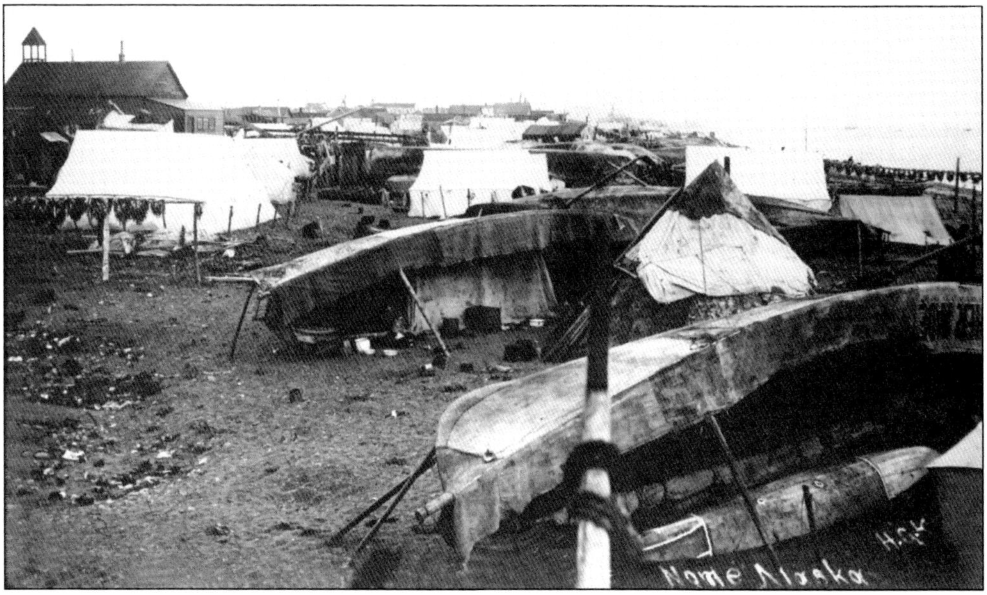

Upon arrival at Nome for the summer, Alaska Native families pitched tents or propped up skin boats (*umiat*) with paddles and driftwood to create lean-to shelters. Additional walrus skins or canvas could be used to cover the open sides of skin boats during high winds. (Photograph by H.G. Kaiser; Carrie M. McLain Memorial Museum, McLain-626.)

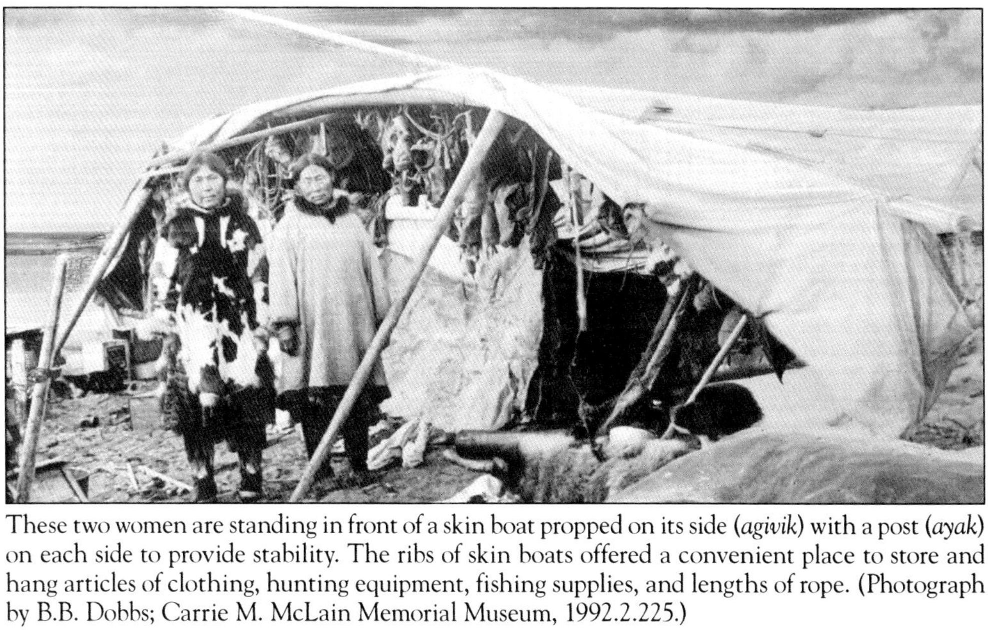

These two women are standing in front of a skin boat propped on its side (*agivik*) with a post (*ayak*) on each side to provide stability. The ribs of skin boats offered a convenient place to store and hang articles of clothing, hunting equipment, fishing supplies, and lengths of rope. (Photograph by B.B. Dobbs; Carrie M. McLain Memorial Museum, 1992.2.225.)

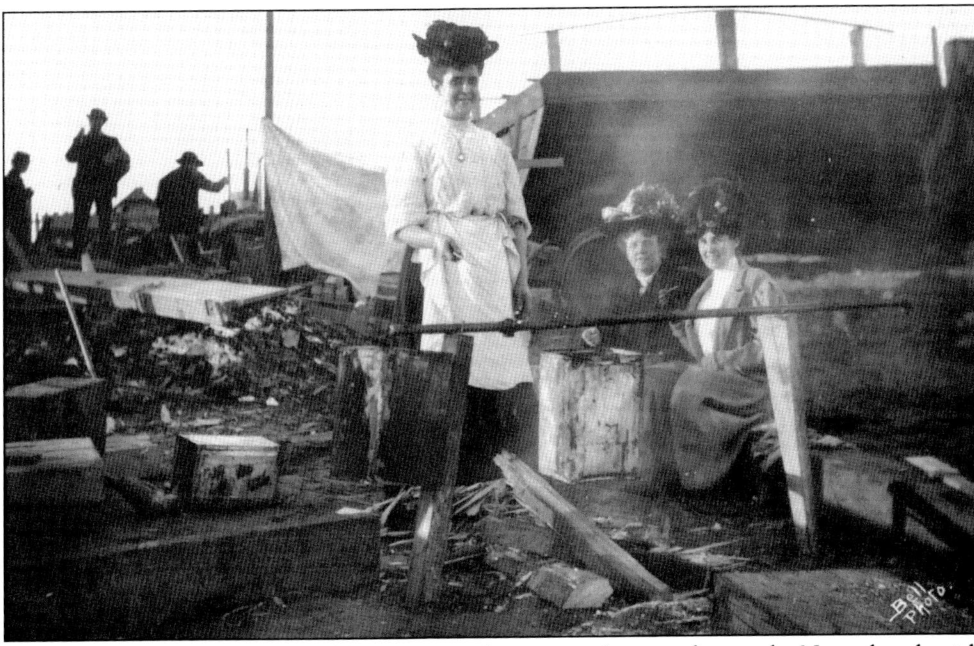

This photograph shows fashionable women cooking around a campfire on the Nome beach with a cookpot made from a tin can between 1903 and 1904. Miners strapped for cash might prepare their own meals outside rather than shell out a hard-earned dollar for a plate of ham and eggs and a cup of coffee at a local restaurant. (Photograph by Arthur L. Bell; Carrie M. McLain Memorial Museum, Bell-SP-863.)

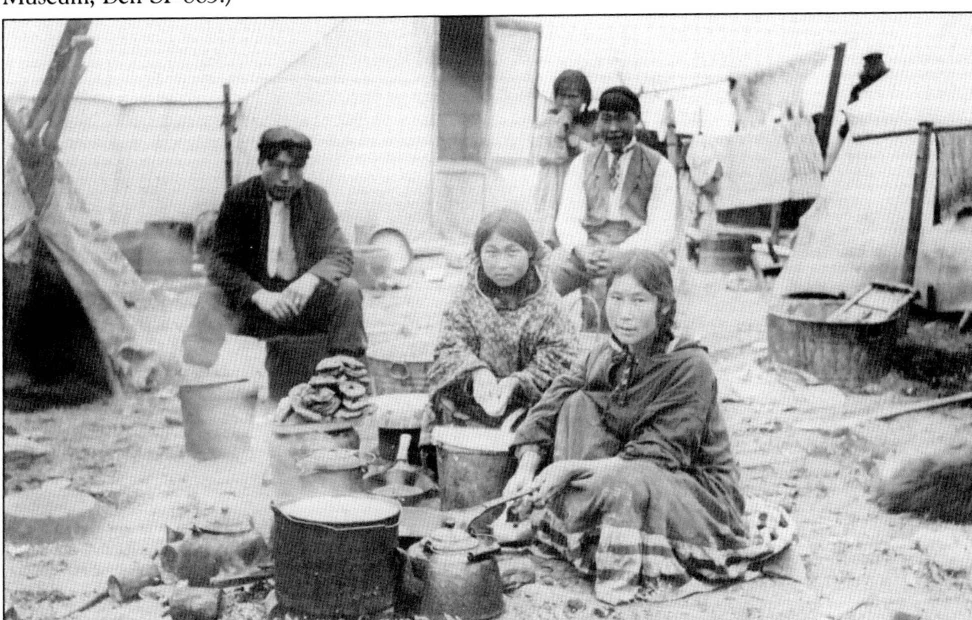

Young women and men from St. Lawrence Island fry up a large batch of donuts over a campfire on the Snake River Sandspit in 1904. The Sandspit offered driftwood for cooking and easy access to the water for fishing and traveling. (Photograph by H.G. Kaiser; Carrie M. McLain Memorial Museum, McLain-742.)

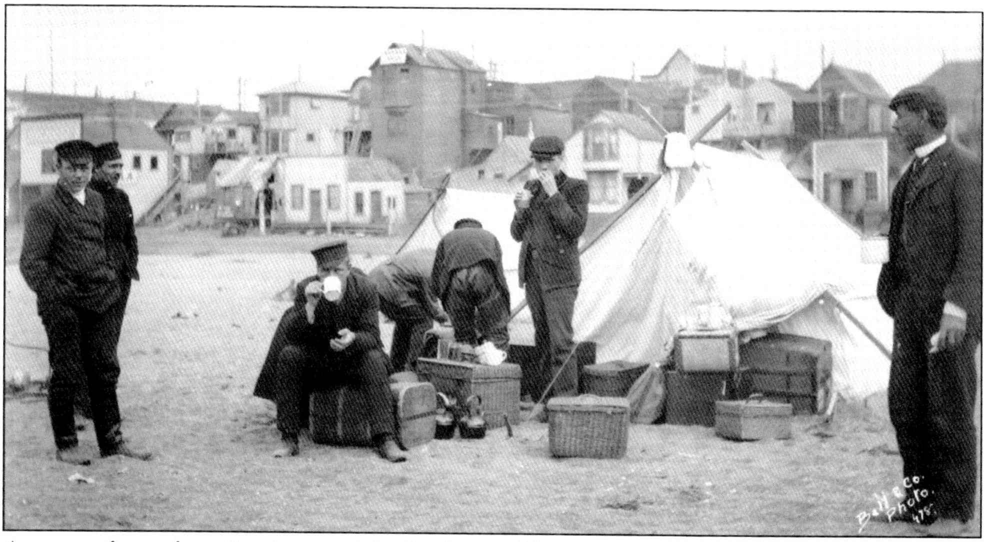

A group of men from Russia use wicker baskets and boxes as temporary tables and chairs during afternoon tea on the Sandspit in 1909. The SS *Victoria* arrived in Nome on June 12, 1909, carrying over 300 people traveling from Russia. Custom officials deported many of the foreign arrivals on the next outbound steamer, while some of the men were able to secure employment with large mining companies in Nome. (Photograph by Arthur L. Bell; Carrie M. McLain Memorial Museum, Bell-N-603.)

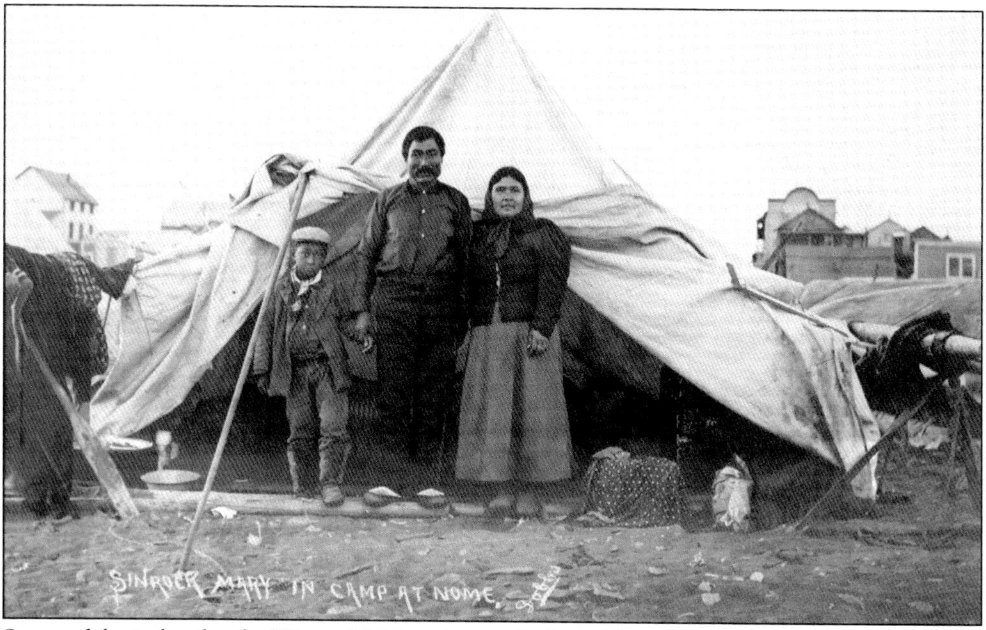

Successful reindeer herder "Sinrock" Mary Antisarlook stands with her second husband, Andrew Andrewuk, and a child outside their tent on the Sandspit in 1902. Sinrock Mary inherited a herd of 500 reindeer when her first husband, Charley Antisarlook, died in the 1900 measles epidemic. She created a thriving business by supplying reindeer meat for Fort Davis and selling to local stores in Nome. Sinrock Mary eventually moved her reindeer to Unalakleet, where it became the largest herd in Alaska and earned her the title "Reindeer Queen." (Photograph by B.B. Dobbs; Carrie M. McLain Memorial Museum, McLain-775.)

Three

Gold Mining

The Nome area contains placer gold that has eroded away from the bedrock and left deposits of nuggets, pickers, and flakes. Gold miners made the first strikes inland along the creeks and rivers, but it was the beach gold that prompted men and women to desert their former occupations and head to Nome. The shorefront remained public property, and any miner could carve out a digging spot about the length of a No. 2 shovel.

The Nome beach held gold for almost 30 miles along the coast, but the richest layers of ruby sand were found just west of the city limits. The shallow placers could be collected with relatively simple equipment, including a rocker box, long tom, or sluice box, which used water and motion to separate the gold from sand and gravel. An industrious miner could make between $15 to $100 a day using a rocker, and by the close of the first year, between $1 and $2 million had been mined from the beach sands.

Although the golden beaches made Nome famous, it was the strikes farther inland along the creeks and valleys that kept the town running. The most valuable claims were located on Anvil Creek, Glacier Creek, Ophir Creek, and Dexter Creek. Anvil Creek produced close to 250,000 ounces of gold, which was worth around $5 million, during its first five years of production. In 1904, J.C. Brown discovered the rich Third Beach Line, which produced between $3 and $4 million in the first two seasons and boosted the population of Nome back up to 12,000 people.

The first bucket-line gold dredge began operation on the Snake River in 1899. By 1912, over 30 gold dredges were mining creeks and rivers on the Seward Peninsula. The gold-mining industry in Alaska collapsed during World War I, and gold production in Nome declined by two thirds during 1916–1918. A new era of mining began in Nome during the 1920s with the introduction of electric-powered dredges that could turn over ground thawed by cool water. Today, the names of historical dredges such as Swanberg Dredge, Dredge No. 5, and the offshore mining dredge BIMA carry a sense of nostalgia for many Nome residents. The majority of mining now takes place offshore, where contemporary prospectors operate floating dredges that use vacuums, or giant excavators equipped with buckets, to turn over material from the Bering Sea floor.

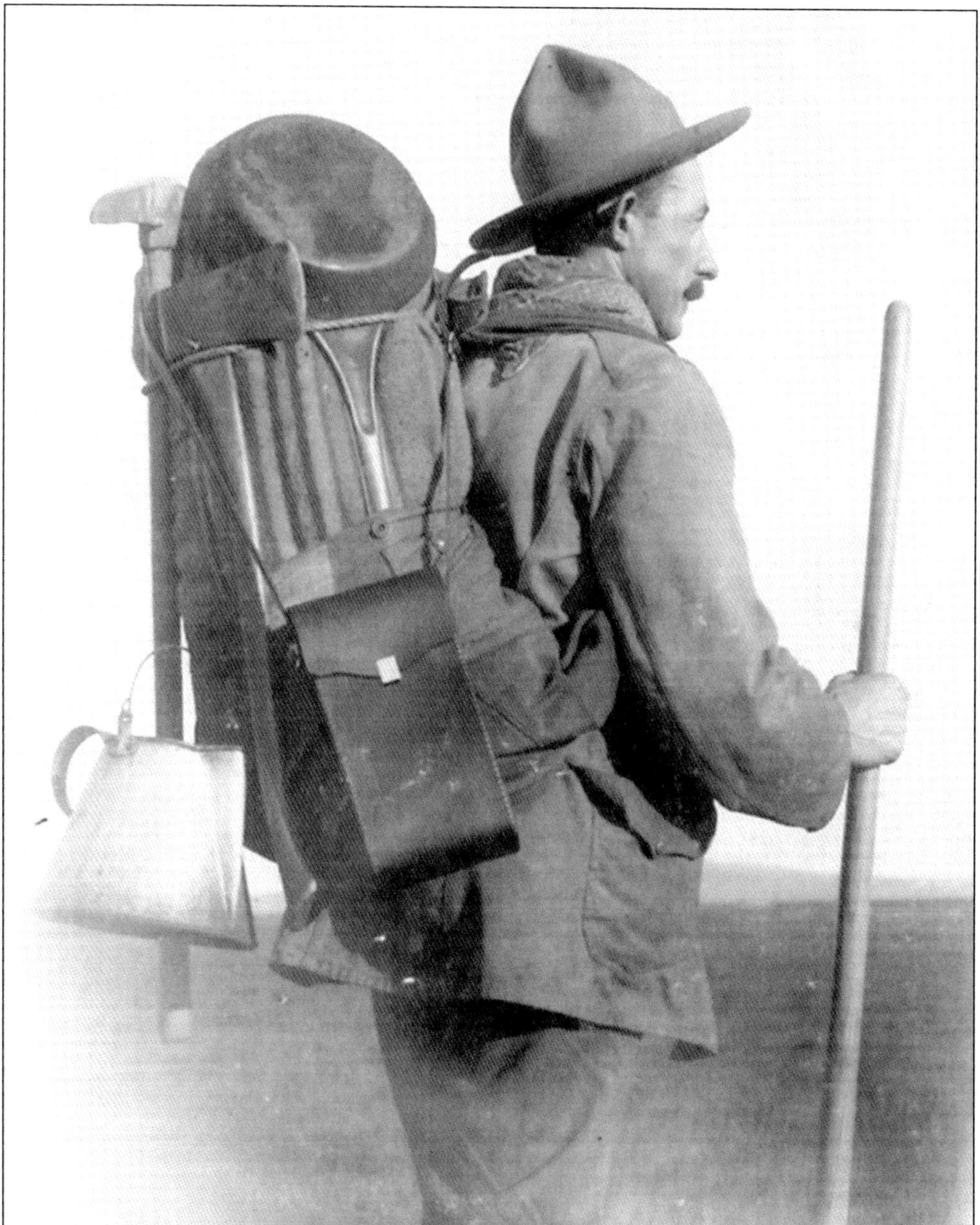

Gold miner Charley Voiyt heads out across the Nome tundra with a pickaxe, kettle, cast-iron pan, splitting axe, and folding-bed camera around 1900. A shovel doubles as his walking stick. Dozens of prospectors traveled by foot over the swampy tundra and hills carrying their packs and sleeping on the damp ground as they made their way to promising claims. Those arriving in Nome with enough money in their pockets could hire the W.J. Rowe Transfer Company to transport their supplies or purchase tickets on the Wild Goose Railroad to reach mining claims in the interior. (Carrie M. McLain Memorial Museum, Rolando-140.)

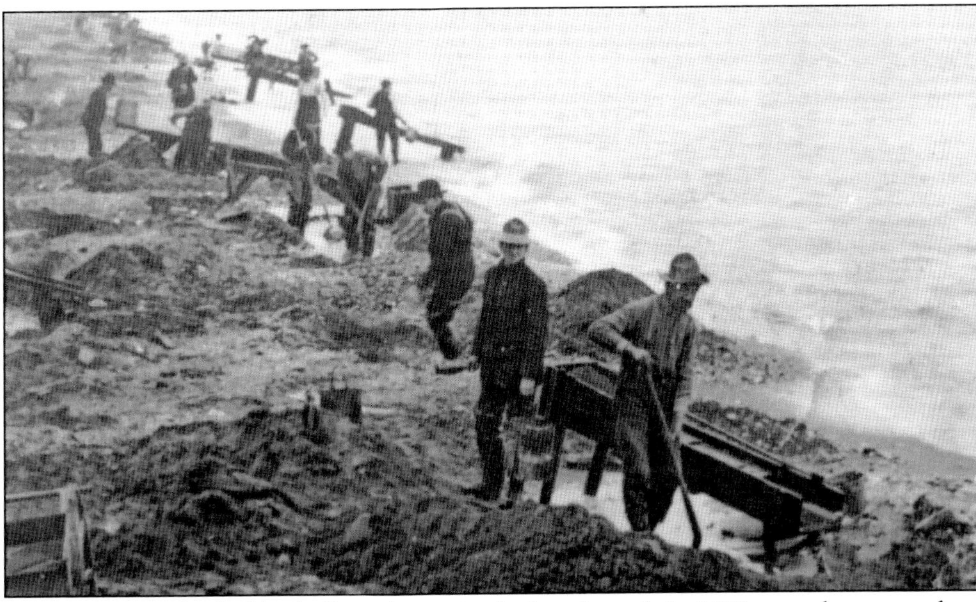

The discovery of valuable shallow placers in the beach sands in 1899 caused a stampede of hopeful miners from the surrounding creeks and the tent city of Nome. Before the close of the season, 500 men had extracted no less than $2 million of gold dust from the sands. The following year saw 8,000 men and women feverishly working shoulder to shoulder on the beach with rockers, long toms, and sluice boxes. (Photograph by H.G. Kaiser; Carrie M. McLain Memorial Museum, NMP-20-716.)

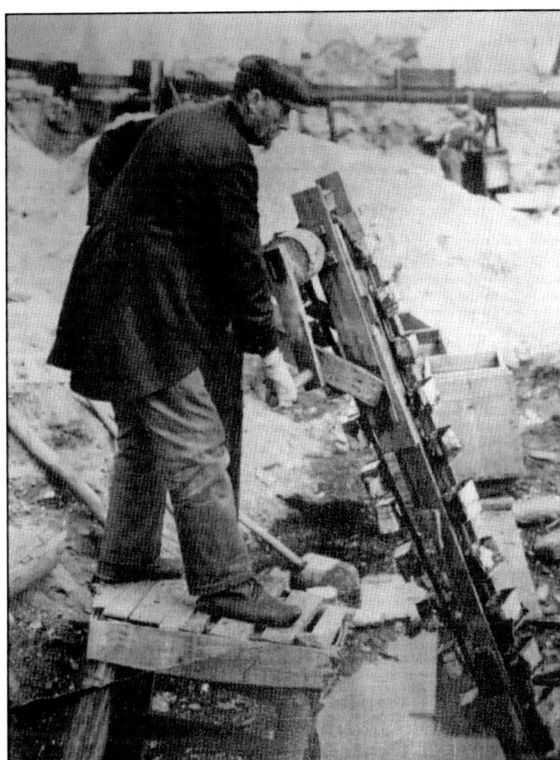

Miners displayed endless creativity in their various attempts to retrieve gold from the beach. This miner is cranking the handle on a bucket line made from empty food tins in 1900 or 1901. Other miners constructed dredges with giant wheels to mine the ocean floor or attempted to power water pumps with homemade treadmills and windmills. (Carrie M. McLain Memorial Museum, NMP-83-5-94.)

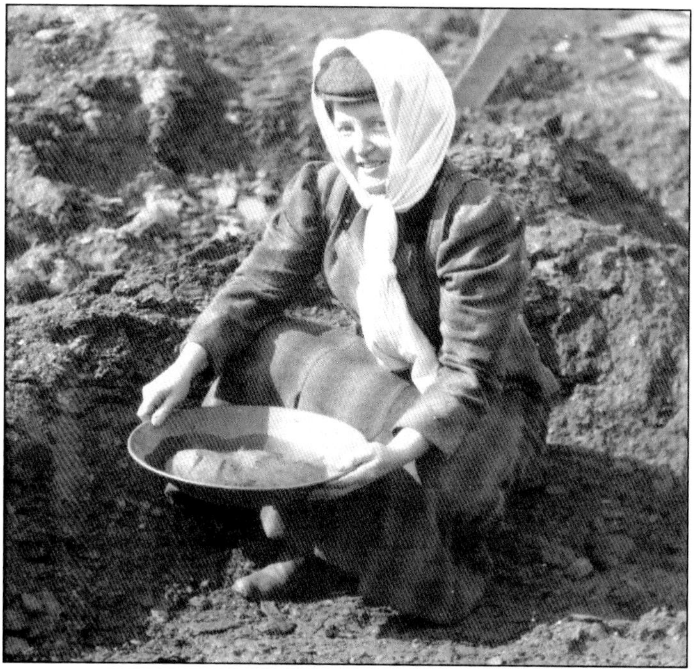

A miner holds a pan of gold dust worth $2,000 from the Bid Six Mining Company at Center Creek sometime between 1903 and 1909. Women in Nome were not opposed to tucking up their long skirts and working on the beach sands right next to the men. (Photograph by Arthur L. Bell; Carrie M. McLain Memorial Museum, Bell-N-578.)

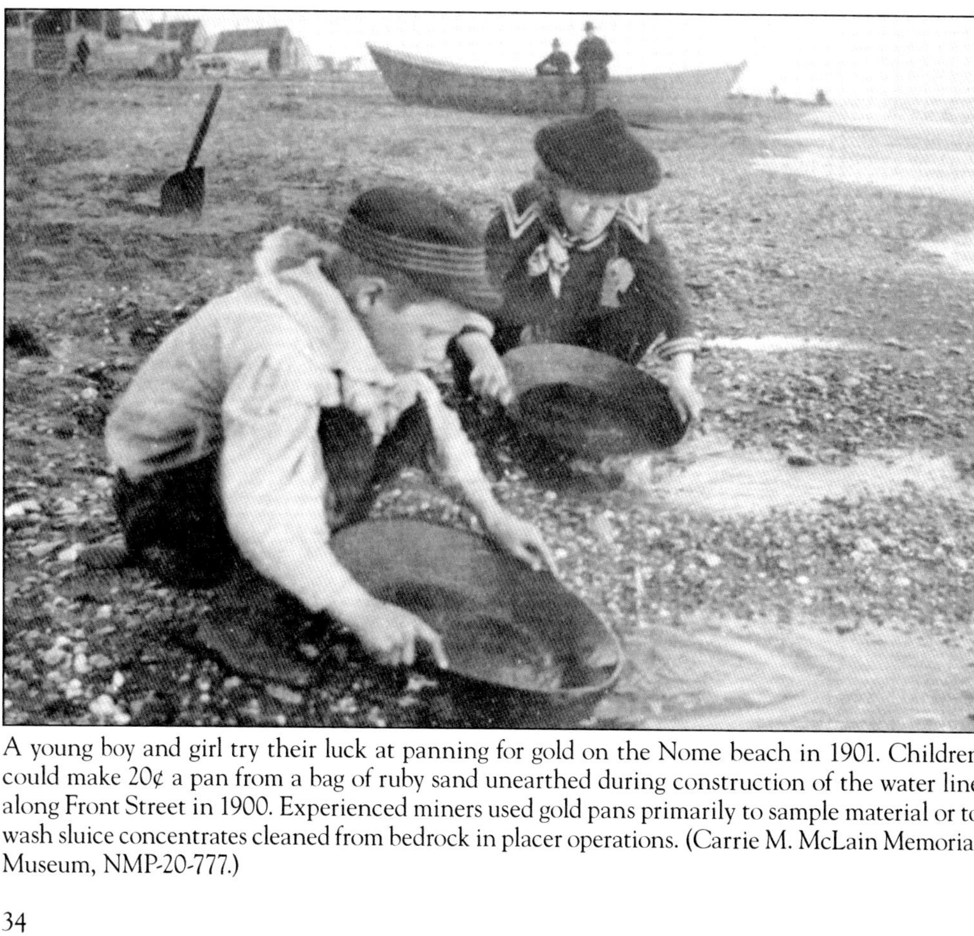

A young boy and girl try their luck at panning for gold on the Nome beach in 1901. Children could make 20¢ a pan from a bag of ruby sand unearthed during construction of the water line along Front Street in 1900. Experienced miners used gold pans primarily to sample material or to wash sluice concentrates cleaned from bedrock in placer operations. (Carrie M. McLain Memorial Museum, NMP-20-777.)

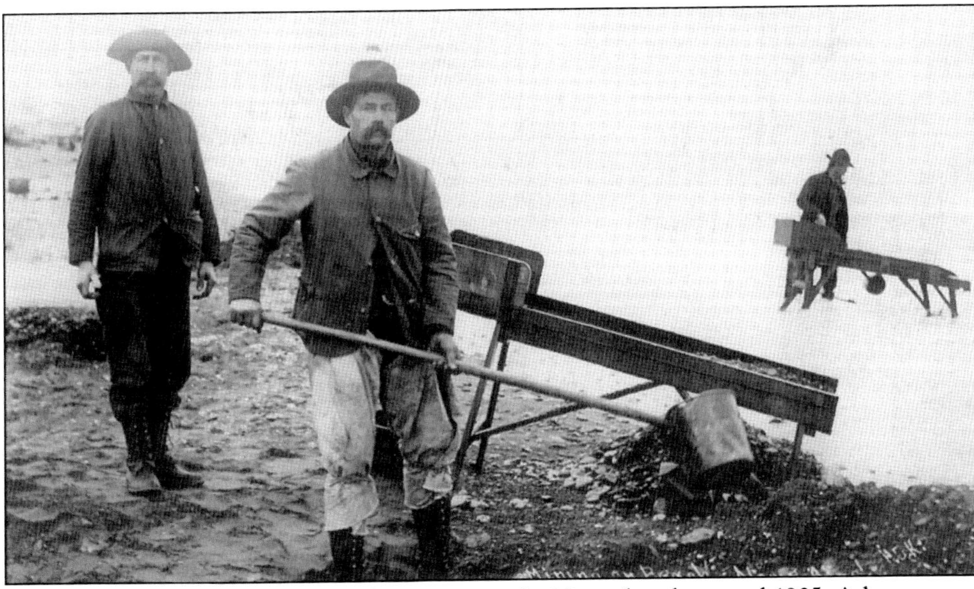

Miners shovel sand and water into long toms on the Nome beach around 1905. A long tom or surf washer was best operated by four miners: one to shovel in sand and gravel, one to maintain water flow, one to remove oversized material, and one to shovel away the tailings. (Photograph by H.G. Kaiser; Carrie M. McLain Memorial Museum, Rolando-125.)

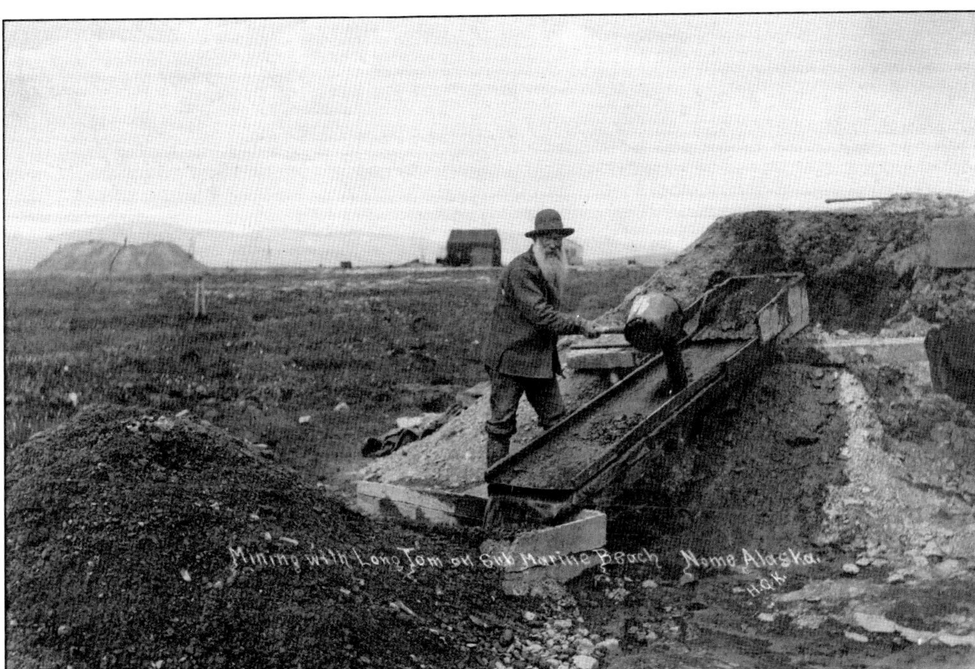

A timeworn prospector is using a long tom on the Sub Marine Beach outside Nome to clean up the remains of a pitched stockpile of pay dirt excavated during the winter from bedrock around 1905. These conical remains were often left behind by large mining companies that operated inland on the Seward Peninsula. (Photograph by H.G. Kaiser; Carrie M. McLain Memorial Museum, McLain-1052.)

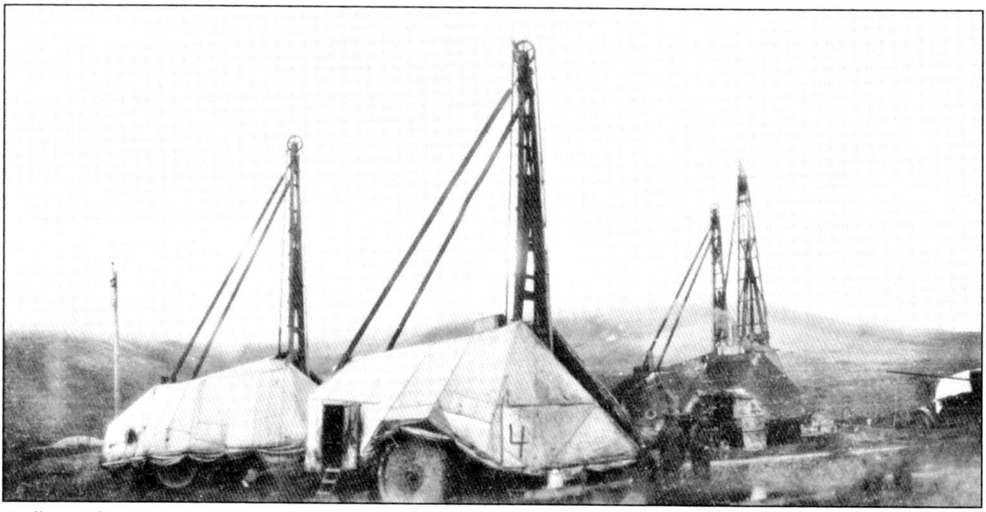

A fleet of Keystone Drills is putting in holes for thawing the permafrost in Nome around 1905. Once the ground was thawed, miners would shovel dirt into a bucket, which was loaded onto a cart and hauled up a line, where the dirt was tipped onto a stockpile that would be worked in the spring. (Carrie M. McLain Memorial Museum, NMP-82-37-89.)

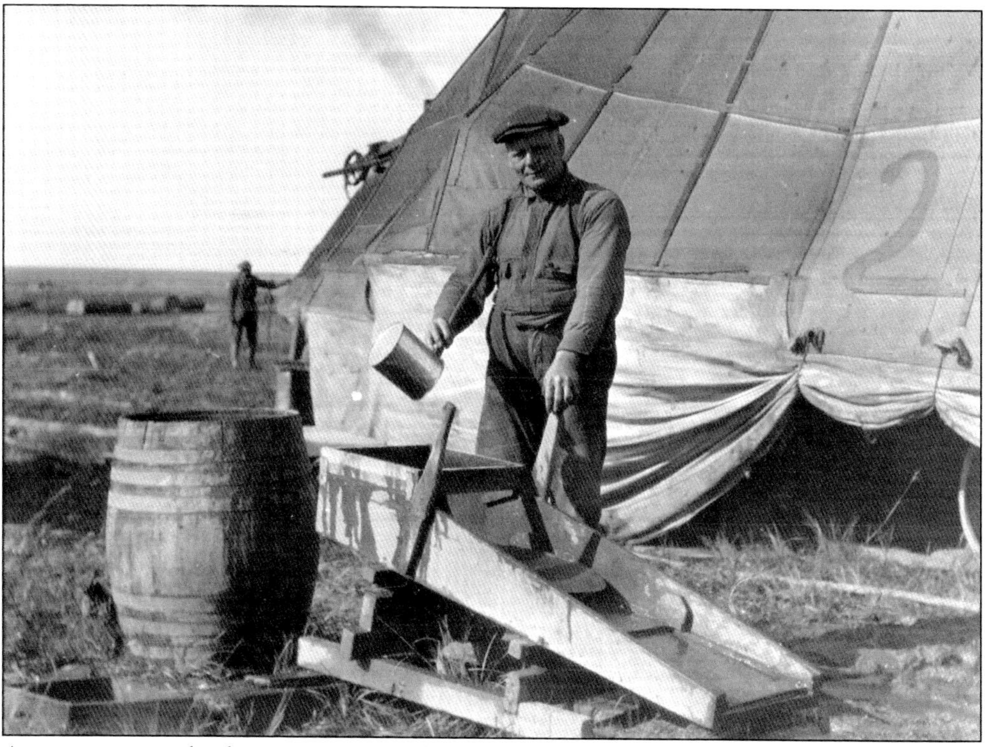

A miner uses a rocker box to process pay dirt next to Keystone Drill No. 2 in Nome around 1905. The rocker box enabled a miner to process more placer material than by simple panning. Sand or dirt was shoveled into the hopper, water was added, and the cradle was rocked from side to side in order to sift the material onto the apron below. As the material and water moved along, heavier minerals like gold would be impeded by raised slats called riffles and collected by hand. (Carrie M. McLain Memorial Museum, 1998.1.48.)

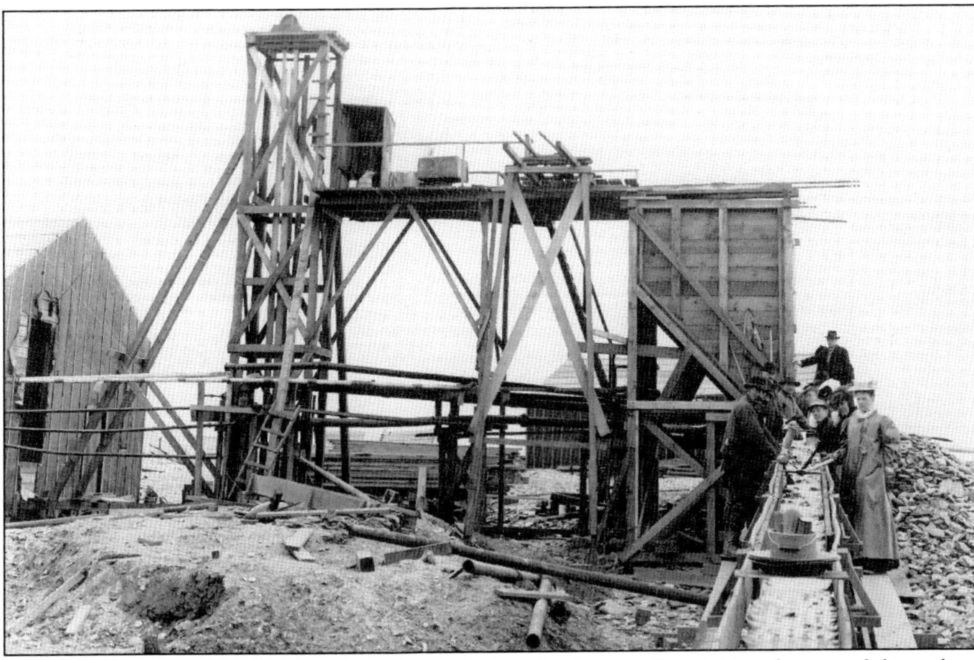

Men and women work a spring cleanup in 1907 on the famous Bessie Bench, one of the richest claims on the Seward Peninsula, with a three-inch-thick paystreak of ruby sand bearing visible gold. During cleanup, pay dirt was shoveled into a line of sluice boxes and washed with water, which pushed the heavy gold to the riffles at the bottom. Dirt and rocks deposited at the end of the sluice boxes were called tailings and were destined to be reworked in later years. (Carrie M. McLain Memorial Museum, 2015.7.)

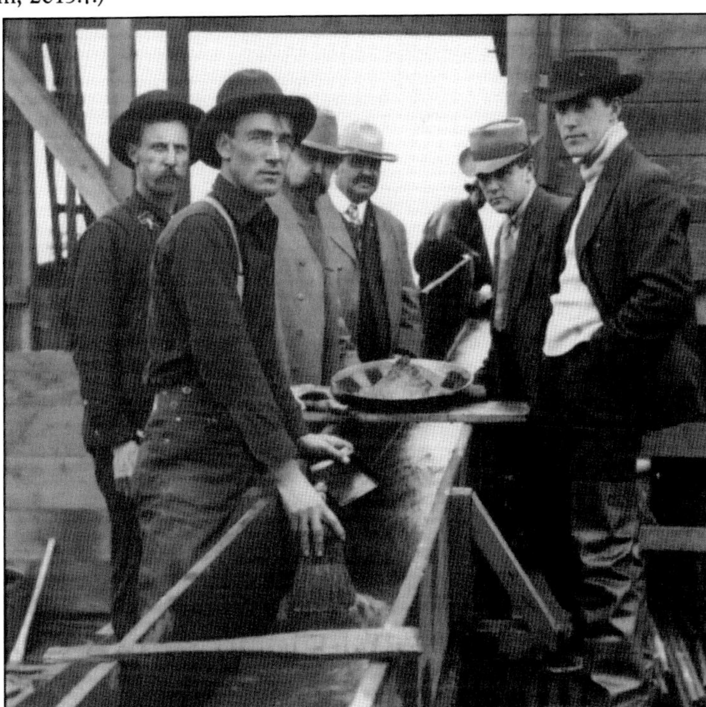

Mining operators pose during a cleanup on No. 8 Cooper Gulch, located on the south side of Anvil Mountain, between 1903 and 1909. Miners routinely stopped the water flow during cleanup to collect the material in the riffles for panning. Final traces of sand and dust would be blown away from gold in the pan. (Photograph by Arthur L. Bell; Carrie M. McLain Memorial Museum, Bell-N-454.)

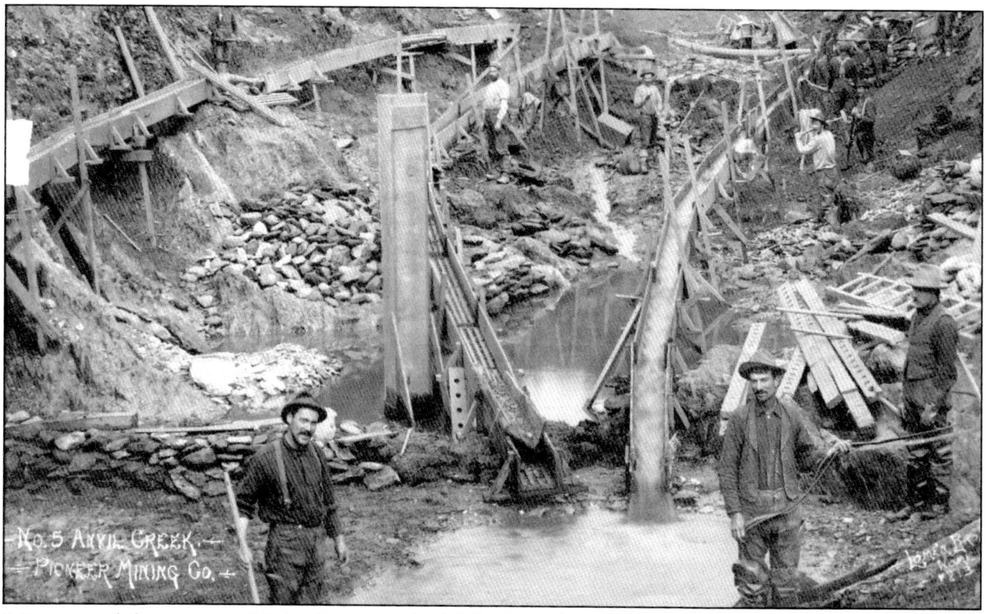

Miners work long rows of sluice boxes at ditch No. 5 on Anvil Creek, owned by the Pioneer Mining Company, in 1910. Removal of placer gold from bench deposits in placer rock required extensive mining operations that diverted water through pipes or wooden plumes into ditches containing rows of sluice boxes that separated the gold and gravel. (Photograph by Lomen Brothers; Carrie M. McLain Memorial Museum, 2015.7.)

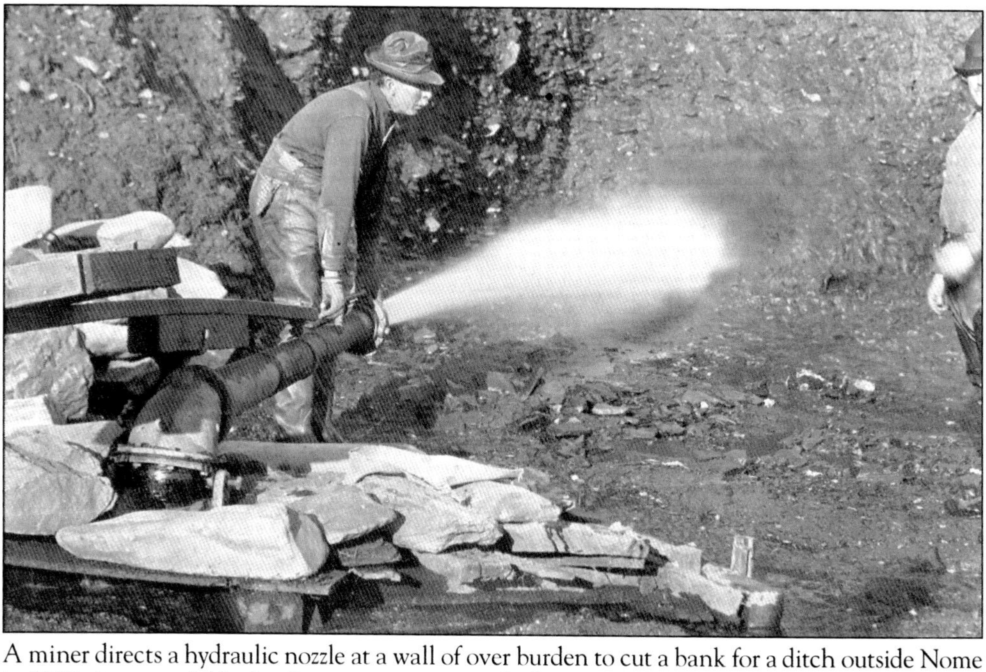

A miner directs a hydraulic nozzle at a wall of over burden to cut a bank for a ditch outside Nome in 1910. A successful hydraulic operation required adequate water pressure, a steep enough grade to carry away the tailings, and a gravel deposit deep enough to make the area profitable to mine. Eighteen sizable ditches were under construction or completed in the Nome mining district by 1905. (Carrie M. McLain Memorial Museum, McLain-1016.)

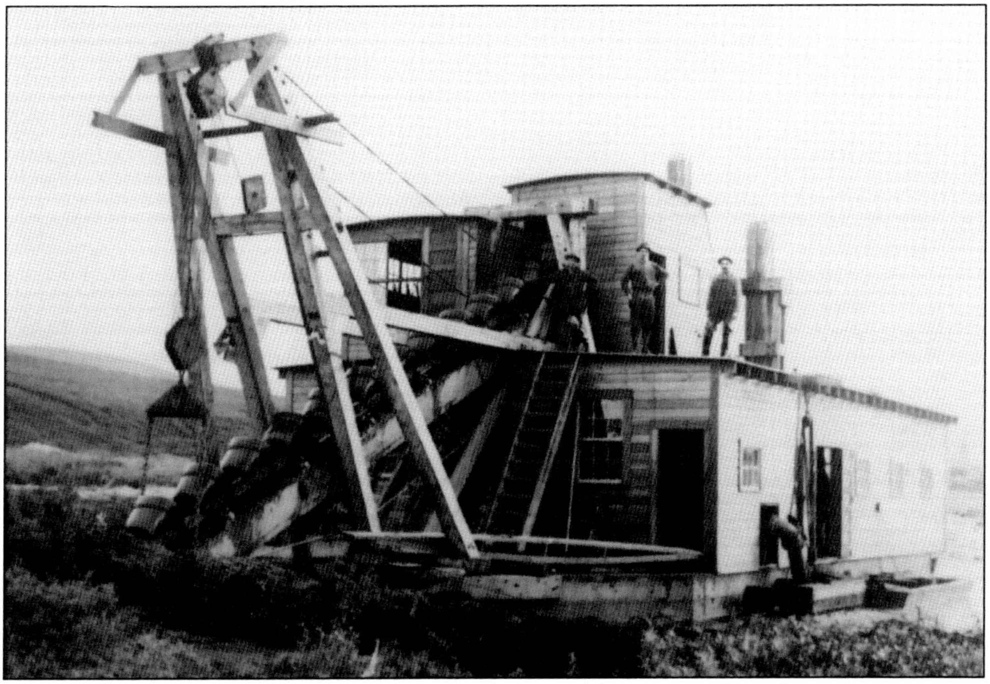

Miners pose on top of the No. 2 Dredge of the Flodin Gold Mining and Dredging Company as it operates on the Solomon River around 1910. Bucket-line dredges carry out four simple operations: digging, classifying material, saving gold, and disposing waste material. An average dredge during the early 1900s could process more material in 24 hours than 2,000 miners. (Photograph by Lomen Brothers; Carrie M. McLain Memorial Museum, NMP-82-25-9.)

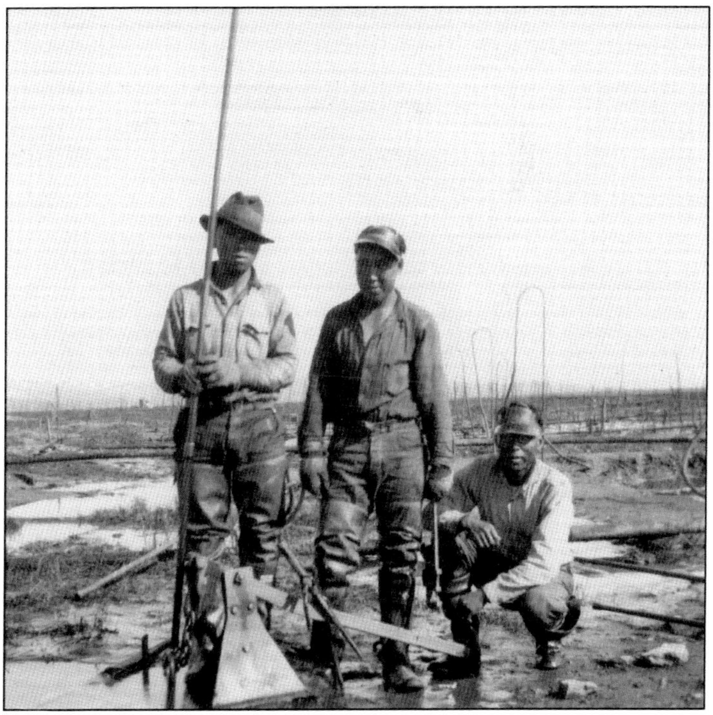

Miners pull metal points from drill bed holes in a muddy thaw field that will be worked by a gold dredge around 1930. Thawing crews sprayed water through the hollow tips of pipes to thaw the permafrost and create a pond for the dredge to move in and dig down to the gold-bearing gravel. (Carrie M. McLain Memorial Museum, 2010.16.)

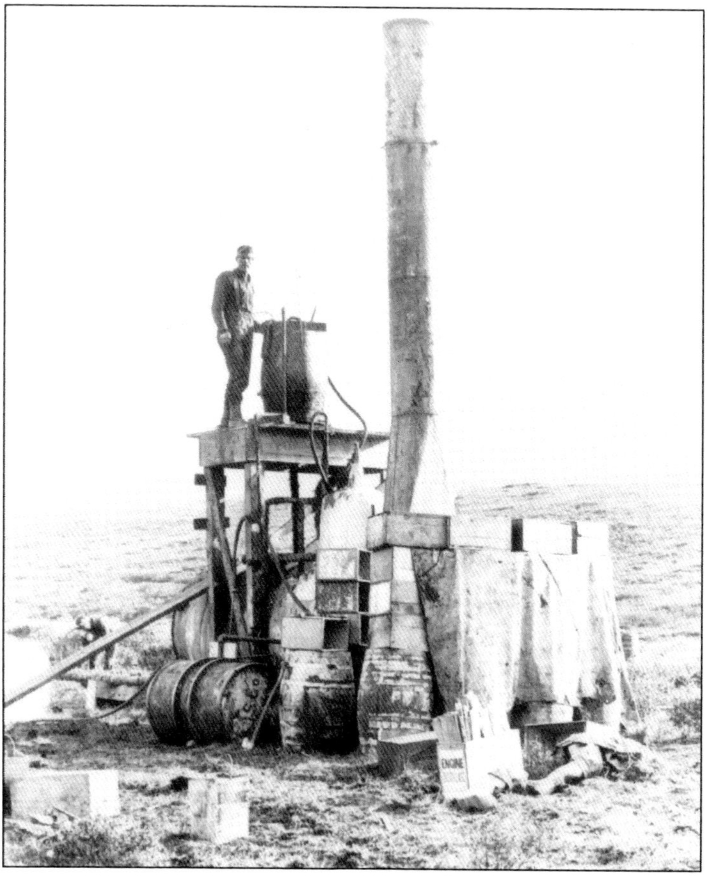

A Nome miner stands on the platform of a homemade steam boiler, which has hoses attached. The boiler forced steam through the hoses and into the frozen ground, where it thawed up to a one-foot layer of permafrost. Miners shoveled the thawed ground into a bucket, hoisted it up, and used the machine again. (Carrie M. McLain Memorial Museum, NMP-20-234.)

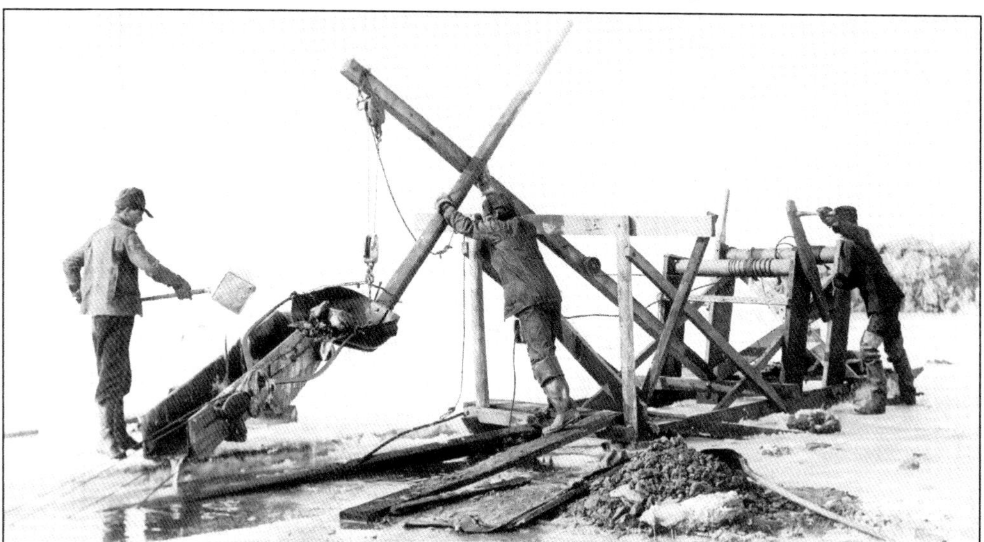

Miners operate a handmade digging machine to reach down through six feet of ice and pull up sand and gravel from the Bering Sea floor around 1905. The bucket of dirt is pivoted to the right and dumped into a long tom, where another miner uses a scoop to wash the material with water and collect the gold. (Carrie M. McLain Memorial Museum, NMP-83-7-1.)

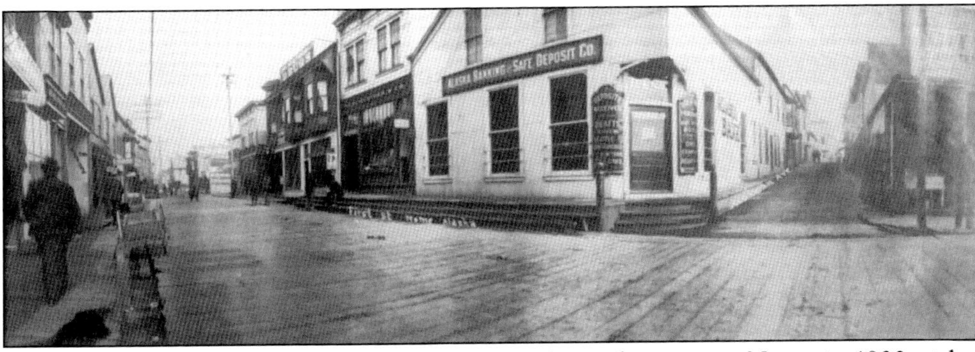

The Alaska Banking and Safe Deposit Company began business in Nome in 1900 with a location at the corner of First Street and Steadman Avenue. The bank dealt exclusively in gold bullion and offered customers the security of fire and burglar proof vaults. Customers made deposits as high as $750,000 during the height of the gold rush. (Carrie M. McLain Memorial Museum, NMP-82-72-202.)

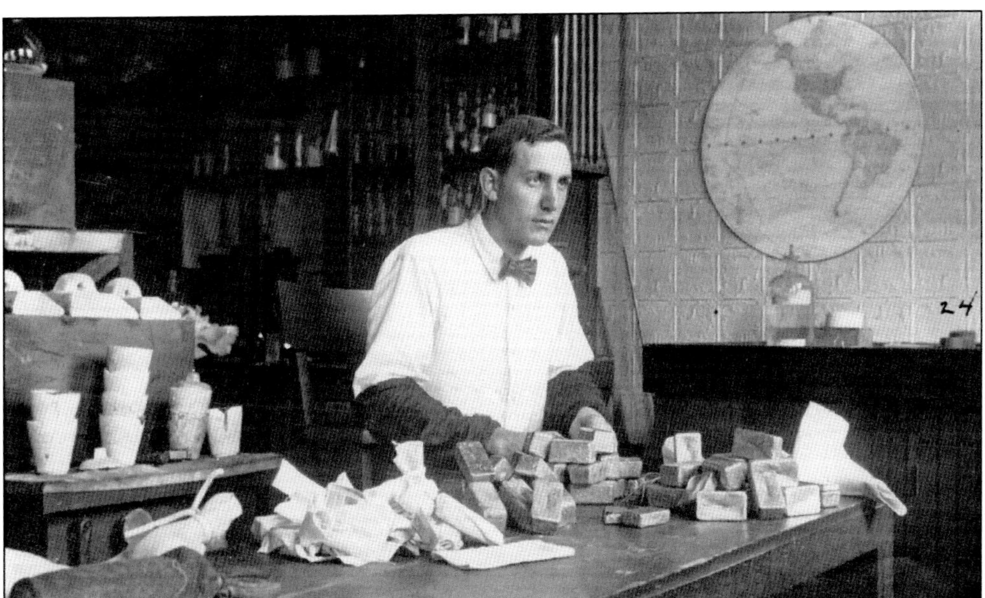

Alfred T. Coston prepares gold bars for shipment inside an assay laboratory in Nome in 1909–1910. Gold miners brought ore to assay laboratories, where experts used chemical solutions to determine the concentration of gold and other valuable minerals in a rock sample. (Carrie M. McLain Memorial Museum, 2001.20.)

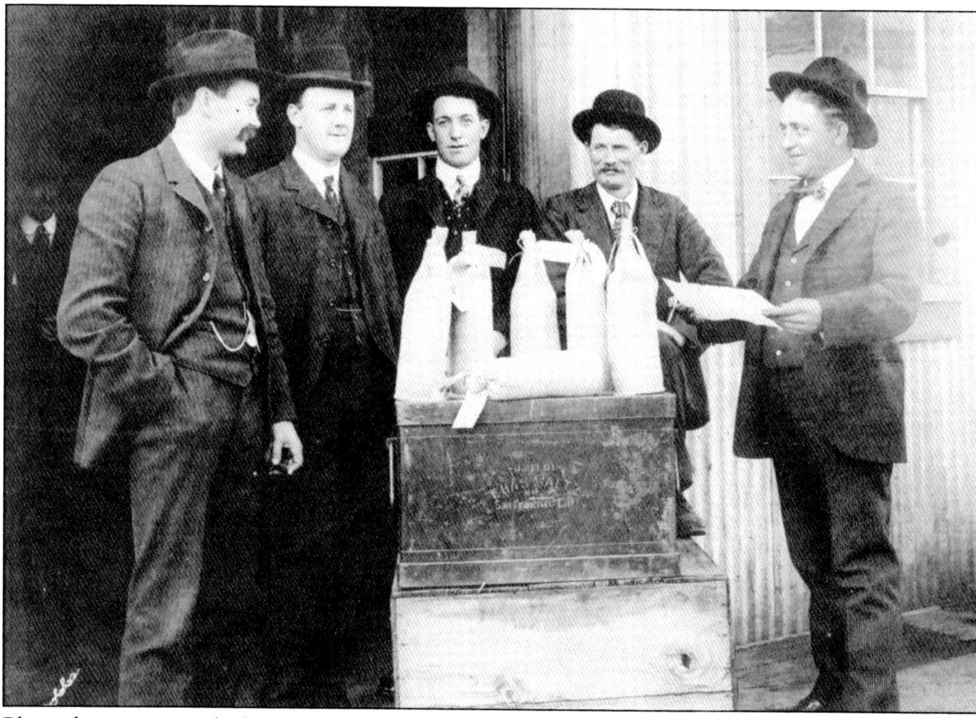

Pleased investors with the Hot Air Bench Mine encircle heavy bags of gold dust and a Waltz safe ready to be shipped from Nome to San Francisco around 1905. The rich claim produced over 300,000 ounces of placer gold and 3,000 ounces of silver by 1903. (Photograph by B.B. Dobbs; Carrie M. McLain Memorial Museum, 96.26.80.)

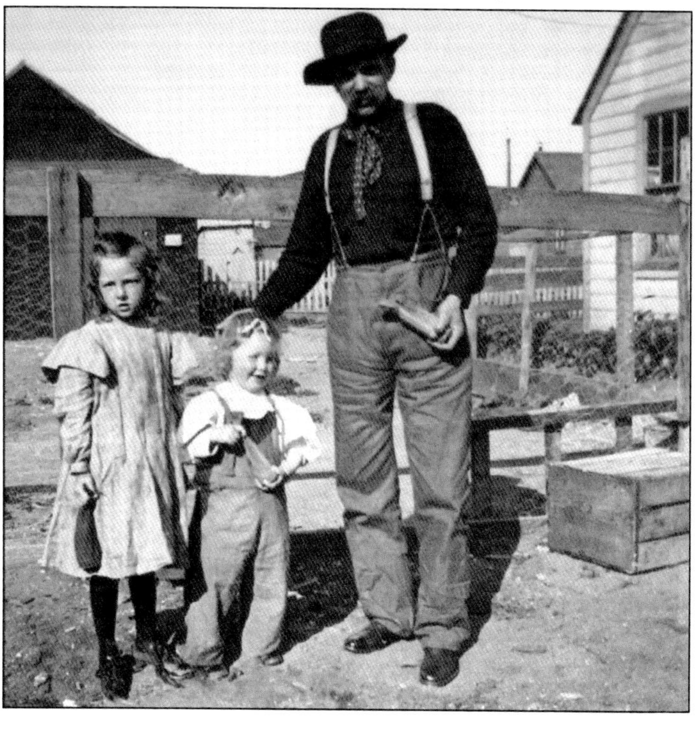

A miner and two young girls hold small bags of gold dust outside a house in Nome. Bags of gold dust were regularly offered as gifts or prizes during the heady days of the Nome gold rush. (Carrie M. McLain Memorial Museum, NMP-11-80.)

Four

Ivory Carving

Open water and summer skies ushered in the arrival of skin boats from surrounding Bering Strait villages to the coast of Nome from the early 1900s to the 1960s. Extended families traveled to summer camps in skin boats over 30 feet long packed with food, clothing, hunting and fishing gear, and often a favorite dog or two. While in Nome, families participated in subsistence and economic activities such as gathering greens and berries, hunting for birds and fish, and selling parkas, mukluks, and ivory carvings.

Bering Strait communities relied on walrus ivory to create whaling and hunting implements, tools, and household articles for over two thousand years. Increased maritime traffic in the Bering Strait during the late 1800s stimulated the creation of an ivory curio market supported by whalers, traders, and a growing number of tourists. The 1898 discovery of gold on Anvil Creek permanently shifted the Western Alaska souvenir trade in ivory carvings from St. Michael, Port Clarence, and Kotzebue to the rapidly growing town of Nome.

Ivory carvers and their families from King Island, Wales, Little Diomede, and St. Lawrence Island set up summer camps on the Snake River Sandspit and carved next to one another under skin boats on the beach. As the population of Nome grew, so did the demand for finely carved cribbage boards, animals, letter openers, salt and pepper shakers, pickle forks, billikens, and models such as miniature shovels, picks, and camp stools. Nome tourists and residents alike eagerly purchased these carvings to decorate their homes, give as gifts, or take back to the Lower 48 as exotic objects from the Arctic.

Carving walrus ivory continues to be an important cultural and economic activity for Alaska Natives in Nome and the surrounding Bering Strait region. Sales of ivory carvings help pay for clothing, heating oil, fuel, and other essential goods. In July 2016, the US Fish and Wildlife Service placed a near-total ban on the commercial trade of African elephant ivory in the United States. Pacific walrus are not an endangered species, and Alaska Native arts and crafts made from walrus ivory are exempt from the domestic ban under the Marine Mammal Protection Act of 1972.

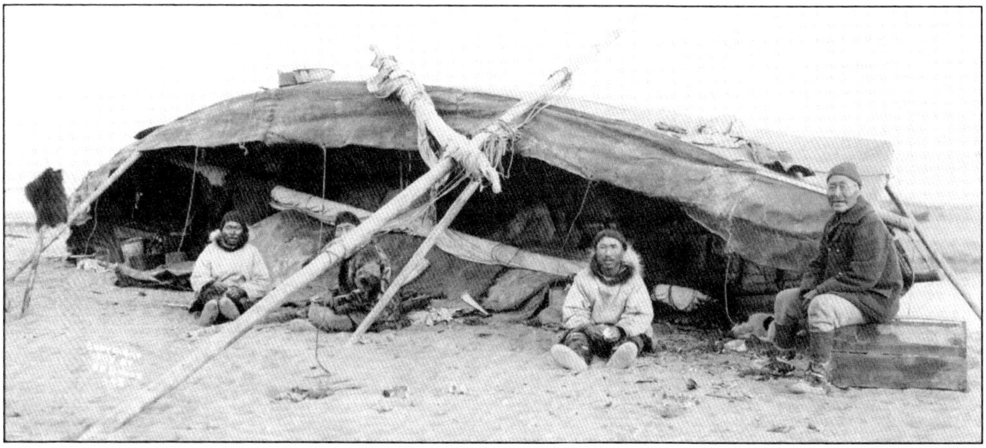

An upturned skin boat provides a windbreak for ivory carvers on the Snake River Sandspit in 1905. Bering Strait carvers discovered a high demand for ivory artwork in Nome, with eager customers who purchased work directly from the carvers on the Sandspit or in local curio shops, such as the Golden Gate Store. (Photograph by B.B. Dobbs; Carrie M. McLain Memorial Museum, 2013.1.68.)

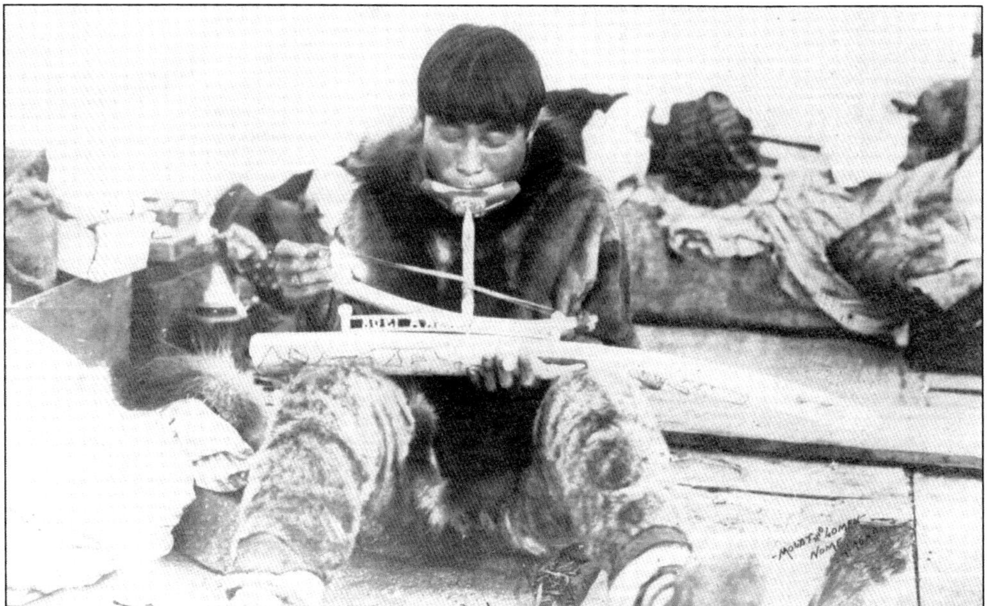

An Inupiat carver uses a bow drill to make holes for an engraved ivory cribbage board (*kaatawik*). Across the top is a delicate ivory rail carved with "Nome Alaska 1908." Cribbage boards can be traced to the 1870s when carvers began engraving large-scale graphics on souvenir walrus tusks, pipes, and cribbage boards at St. Michael on the southeastern coast of Norton Sound. St. Michael served as the headquarters for the Alaska Commercial Company and a gateway to the Alaskan interior via the Yukon River during the gold rush of 1897. (Photograph by Moldt and Lomen; Carrie M. McLain Memorial Museum, 1979.1.491.)

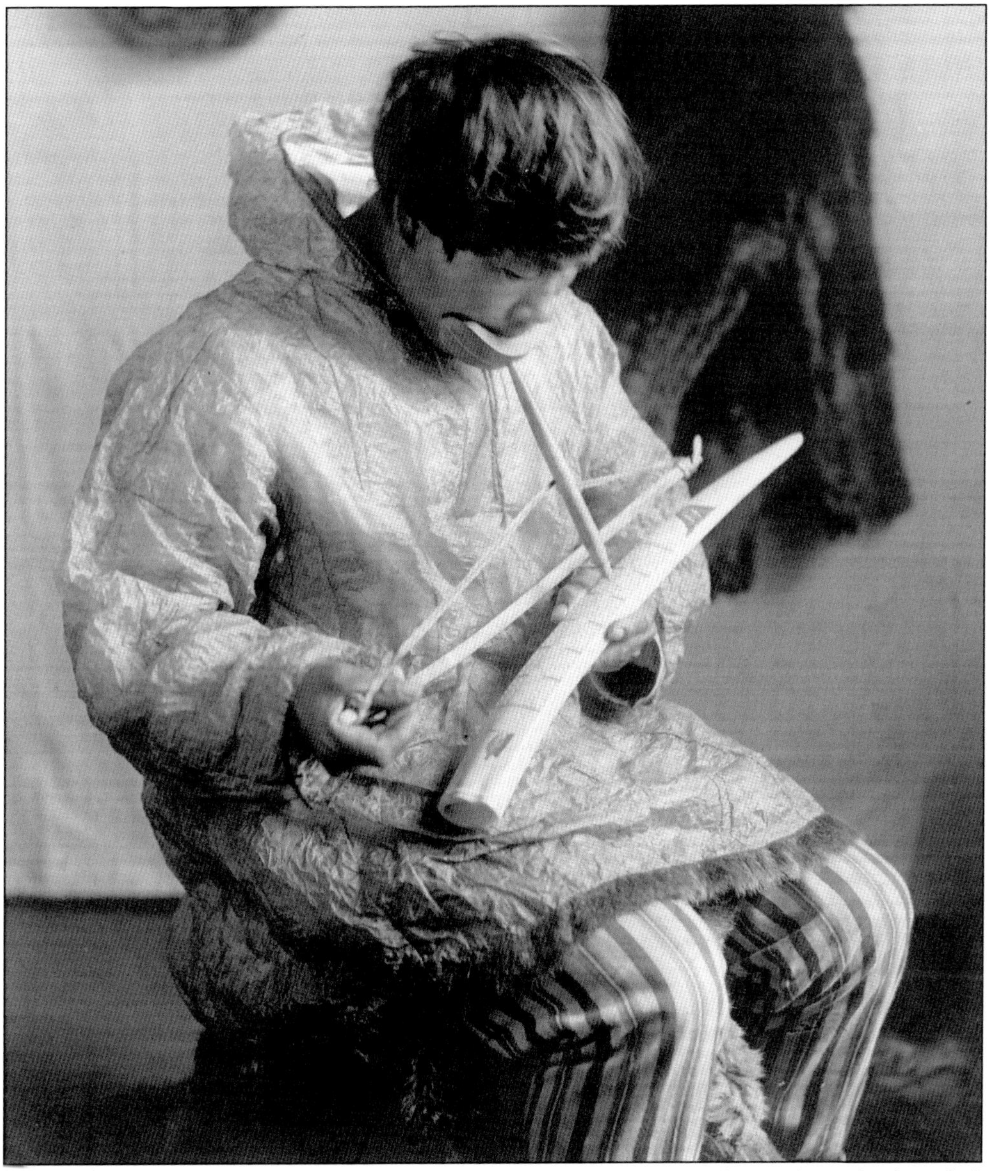

The celebrated ivory carver Angokwazhuk "Happy Jack" poses in a gutskin parka with a bow drill and engraved cribbage board in Nome in 1912. Happy Jack (c. 1870–1918) was born in Ayasayuk, a village near Cape Nome, and grew up on the Diomede Islands where an unfortunate hunting accident cost him his feet and the ability to pursue critical subsistence activities. During the 1890s, Happy Jack sailed aboard two whaling ships where he learned the art of scrimshaw. He arrived in Nome in 1900 and quickly gained a following for his ivory carvings featuring detailed and realistic engravings of dog teams, reindeer herds, village scenes, and portraits. Happy Jack can be considered the first professional Inupiat artist. His work fostered a modern ivory art market that continues today. (Photograph by H.G. Kaiser; Carrie M. McLain Memorial Museum, 2013.1.32.)

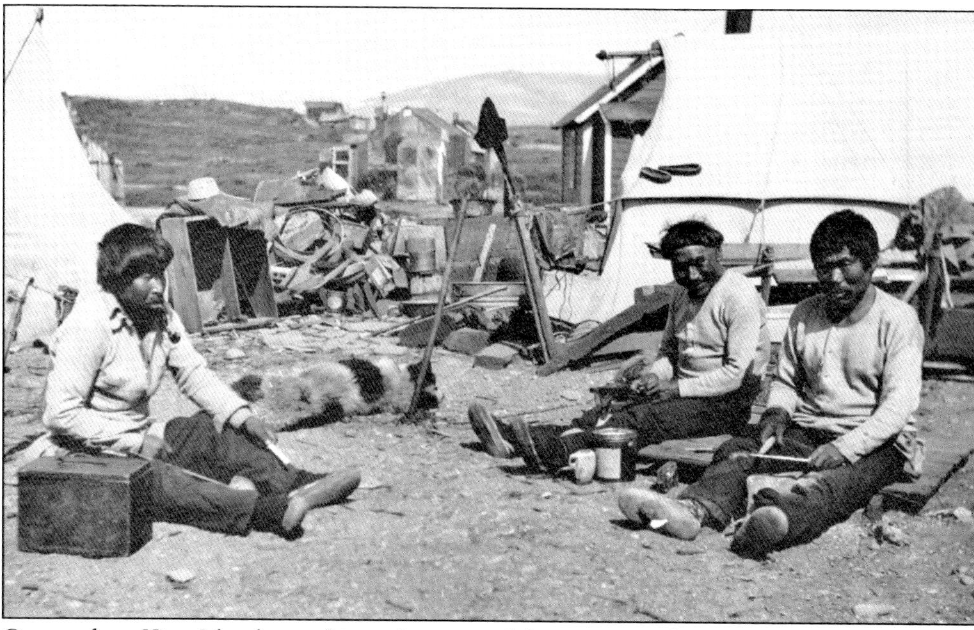

Carvers from King Island are taking advantage of the warm weather to file walrus tusks outside their tents on a sandy beach around 1910. The presence of a pipe and bemused smiles indicate that good conversation is mitigating the strenuous endeavor of carving with hand tools. (Carrie M. McLain Memorial Museum, 2005.11.)

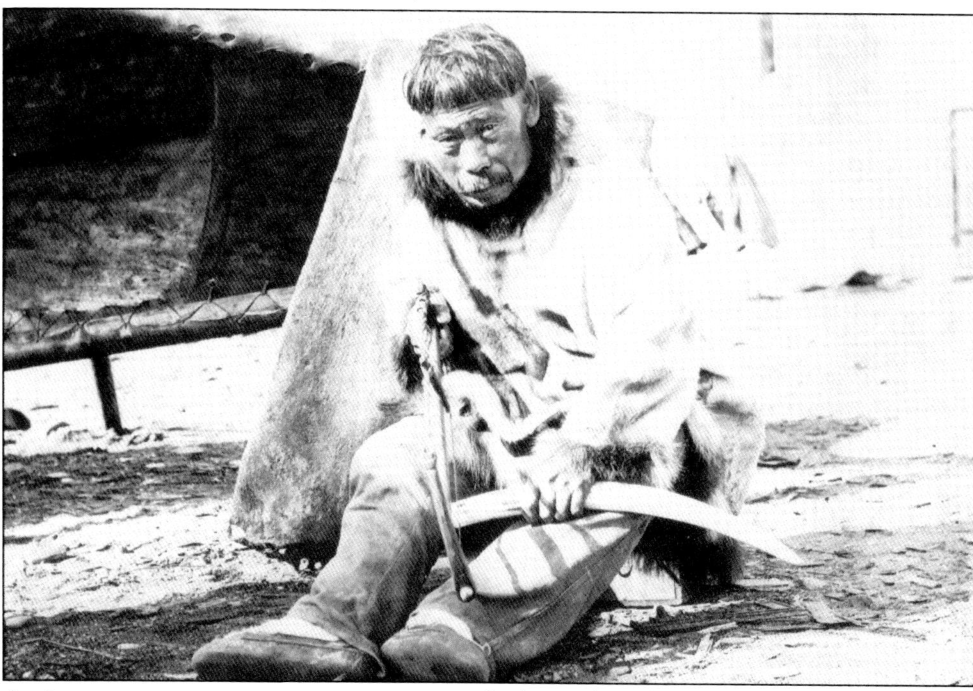

An ivory carver uses a coping saw to remove the base of a walrus tusk in Nome around 1915. Essential carving tools during the early 1900s included an adze, knife, whetstone, bow drill, coping saw, file, and engraver, which were often stored in a bentwood toolbox. (Carrie M. McLain Memorial Museum, 2001.18.)

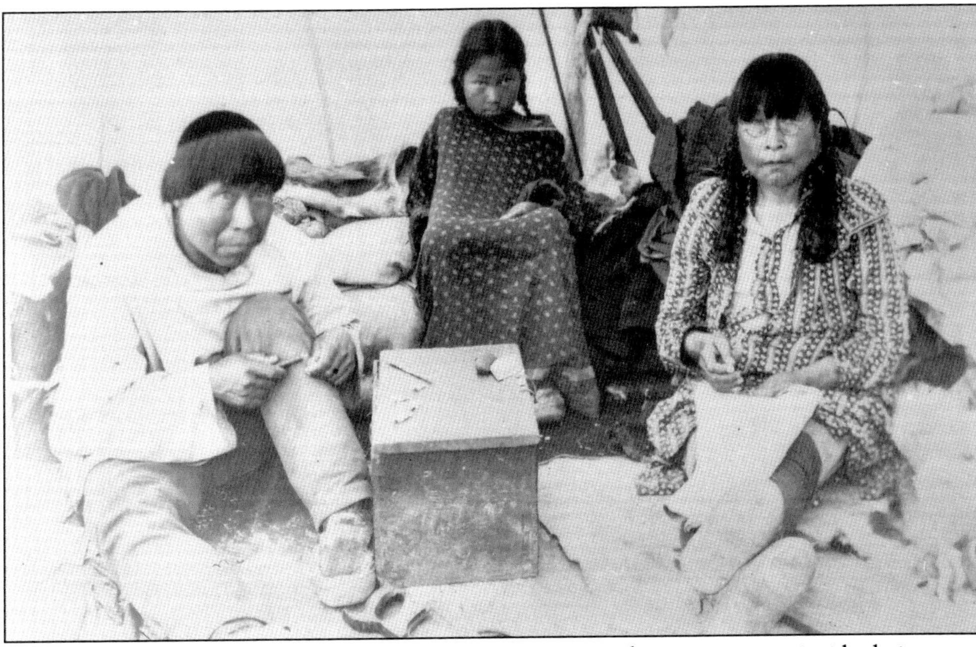

A family from St. Lawrence Island works on ivory carvings and sewing projects inside their tent at Nome around 1905. Fur parkas and reindeer-skin boots provided exceptional warmth for newcomers to Alaska, while both traditional tools and souvenir carvings were displayed as decoration within gold rush homes. (Carrie M. McLain Memorial Museum, NMP-82-72-214.)

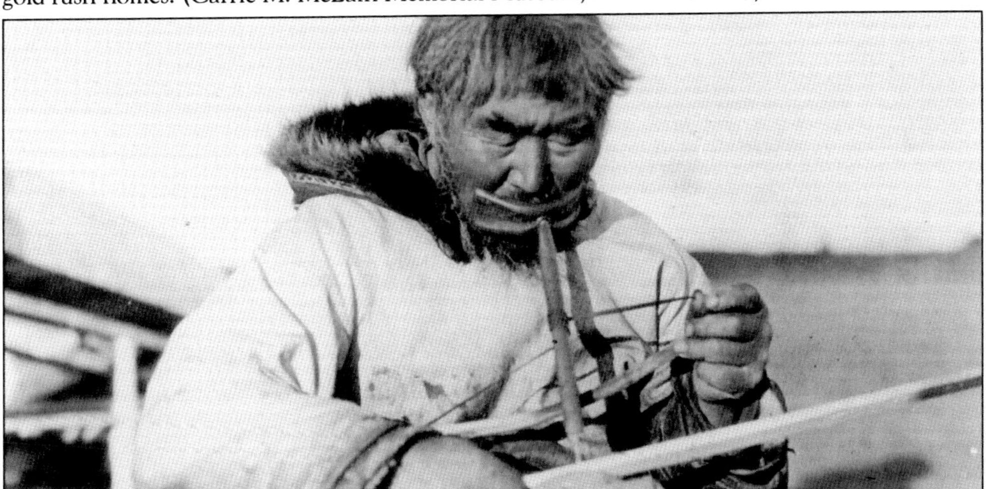

An Inupiat carver demonstrates the use of a bow drill on the Nome beach around 1915. Bow drills are multipurpose tools used to start fires and bore holes in ivory, bone, or wood. The tool consists of an ivory or bone drill bow (*pitiksiaq*), a wood drill fitted with a metal bit (*niuun*), and a mouthpiece with a stone socket (*kiŋmiaq*). (Photograph by Curtis Jacobs; Carrie M. McLain Memorial Museum, McLain-745.)

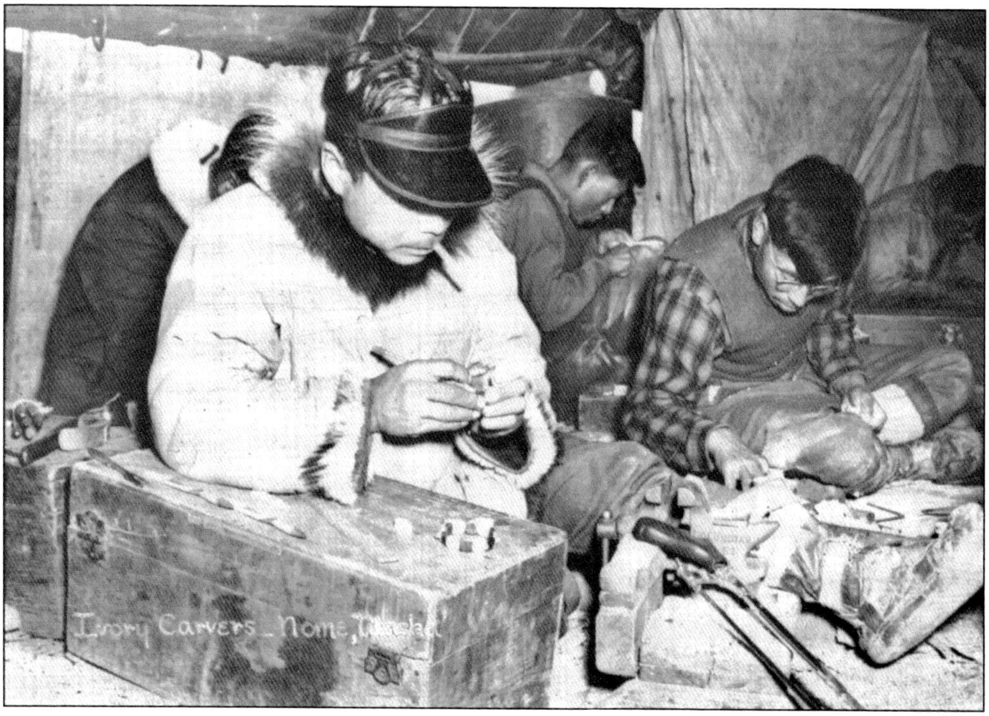

Carvers are using hand tools under a skin boat to fashion ivory bracelets, pickle forks, and letter openers in Nome around 1930. Young people learn to carve through observation of older relatives and gradual participation in the carving process. Bering Strait carvers traditionally worked together in community houses (*qargiit*) during the winter or under skin boats at summer camps. Many King Island families moved to Nome during the 1950s and 1960s, and carvers worked side-by-side in a community Quonset hut located at East End. Today, several carvers take advantage of the shared tools and camaraderie of the carving workshop at the King Island Native Community Building in Nome. (Carrie M. McLain Memorial Museum, 2002.17.1.)

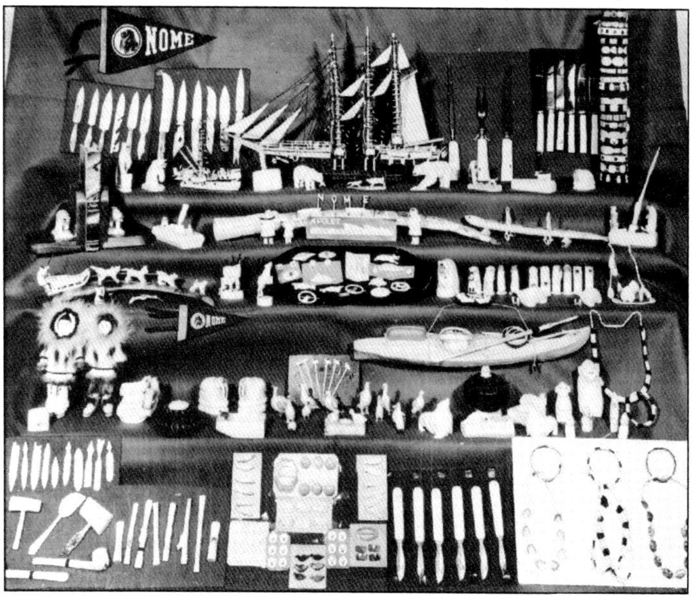

An eye-catching display at the A. Polet store in Nome around 1940 features ivory letter openers, polar bears, bracelets, penholders, earrings, hatpins, billikens, pipes, cigarette holders, buttons, butter knives, and necklaces. Antonio Polet arrived in Nome in 1900 and ran a popular general store for 50 years that sold everything from needles and coal to ivory carvings and fur parkas. (Carrie M. McLain Memorial Museum, McLain-911.)

Five

Building a Town

The gold rush to Nome delivered waves of prospectors, businessmen, and families to the southern coast of the Seward Peninsula. Newcomers discovered that the windswept tundra and barren hills offered few raw materials from which to construct stores and homes. Steamships from Seattle, San Francisco, and Portland quickly responded to the need for lumber with millions of board feet shipped in at an average of eight dollars a ton. Almost overnight, a city of saloons, dance halls, hotels, restaurants, and stores sprung up along the sandy shorefront.

During the summer of 1900, the town of Nome stretched out precariously close to the water's edge in a thin line about a mile long and two blocks wide. The location of the townsite was less than ideal, and streets required continual ditching, draining, and graveling to keep them passable. Almost 20,000 people kicked up dust down the narrow paths while the whistle blasts of ships and whinnying of horse teams added to the clamor of new buildings under construction.

The City of Nome was officially incorporated in April 1901, with the first municipal officers focused on planking the streets and organizing a fire department. Soon the largest town in Alaska could boast of electric lights, piped water, three churches, a public library, and a 50-bed hospital. Over 200 students attended the two public schools located in town and on the Snake River Sandspit. Residents formed a ski club, put on theater productions, and organized citywide celebrations for the Fourth of July with colorful parades and competitions.

Crowded conditions down the planked streets of the business district contributed to the spread of devastating fires in 1901, 1905, and 1934. Between fires, residents were hammered with severe Bering Sea storms in 1902, 1913, 1945–1946, and 1974 that drove ice cakes onto the beach, eroded the waterfront, and crushed buildings. The series of destructive fires and storms significantly altered the development patterns and the built landscape of Nome. The city plat was redrawn with wider roads set farther inland in 1934. In 1951, the federal government constructed a 3,350-foot-long protective seawall of granite rock to help hold back the turbulent surf of the Bering Sea.

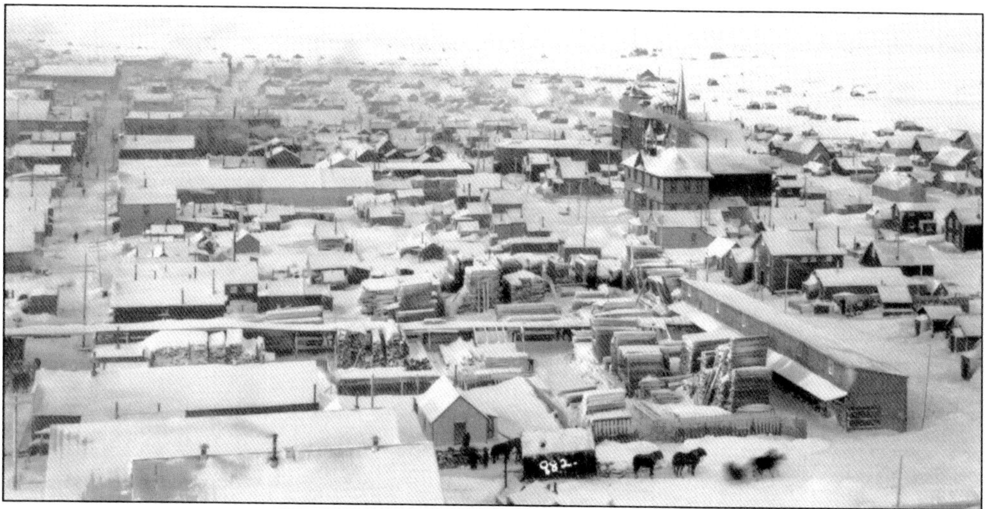

Photographer Arthur L. Bell clambered onto a crane owned by the John J. Sesnon Lighterage Company and used its pivotal arm to take a series of images of gold rush Nome between 1906 and 1909. This view looks west over the town and extends into open tundra. Horses are pulling a small house on runners in the foreground. Thousands of board feet of lumber can be seen stacked in the center. At far left, residents are trudging through deep snow down a path illuminated by electric lights. Tall buildings on the right include the Nome Public School topped by a bell tower (front), St. Joseph Catholic Church (center), and Holy Cross Hospital (rear). (Photograph by Arthur L. Bell; Carrie M. McLain Memorial Museum, Bell-N-129.)

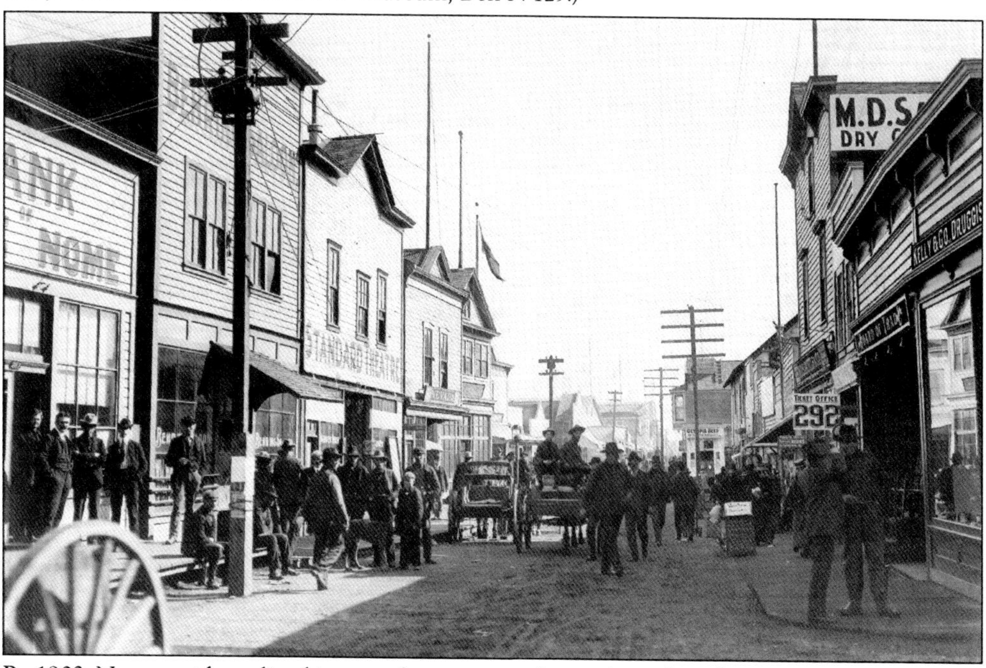

By 1903, Nome residents lived in a modern town with wooden sidewalks, an electric light plant, and telephone and messenger services. Local stores offered all the necessities of life and many of the luxuries found in a big city, as seen in this view down Front Street featuring the Bank of Nome, Behring Saloon, Standard Theatre, Nevada Bar, Kelly & Co. Druggist, Board of Trade saloon, and M.D. Samuels & Co. Dry Cleaners. (Carrie M. McLain Memorial Museum, 96.26.61.)

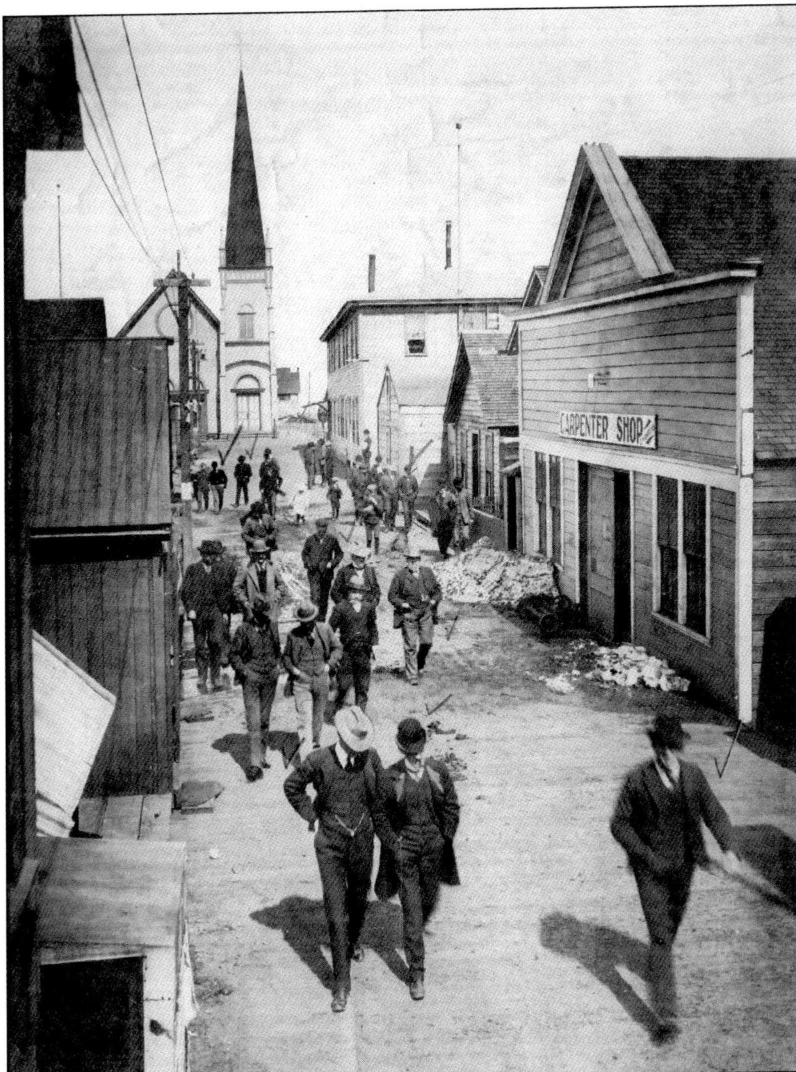

A group of men lean in for discussion after leaving the Nome courthouse on Steadman Street in 1903. The courthouse served up regular sessions of drama beginning in July 1900 when Alexander McKenzie and Arthur H. Noyes, newly appointed district judge, arrived and began a fraudulent scheme to seize many of the richest mining claims in a gambit that came to be known as "The Spoilers." Judge Noyes remained in office until 1902, when he was convicted for contempt of court and replaced by Judge James Wickersham. Past the courthouse at the end of the street rises St. Joseph Catholic Church, built in 1901. The church featured a steeple almost 100 feet tall topped by an electric cross to guide travelers lost out on the tundra or sea. By 1927, high heating costs and a diminishing congregation prompted its closure and a search for a more cost-effective structure. The US Smelting, Refining and Mining Company purchased the church in 1946 and used it as a cold storage facility until 1993, when the building was donated to the City of Nome. In 1995, the Nome Common Council voted to commit city funds to a matching federal grant and restore St. Joseph Catholic Church. The church, now known as Old St. Joe's Hall, was moved to its present location on Bering Street in April 1996. Preservation efforts, including restoration of the interior, creation of a new rose window, and replica of the original steeple, were completed in 1997. (Photograph by Otto D. Goetze; Carrie M. McLain Memorial Museum.)

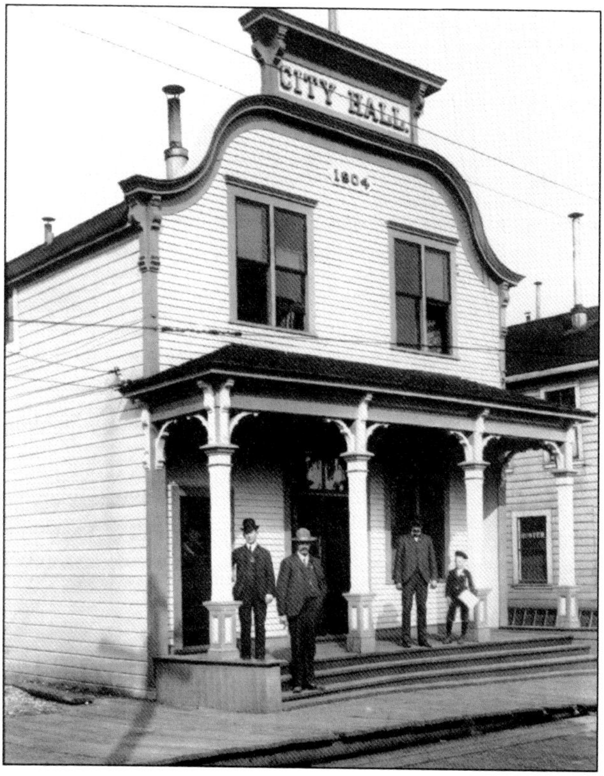

A trio of smartly dressed men and a paperboy peer at the camera from the covered entrance of the former city hall in 1905. Nome elected its first city council in April 1901, and the seven members quickly set out to pass ordinances against gambling, prostitution, and firing guns within city limits. (Photograph by F.H. Nowell; Carrie M. McLain Memorial Museum, NMP-82-4-1.)

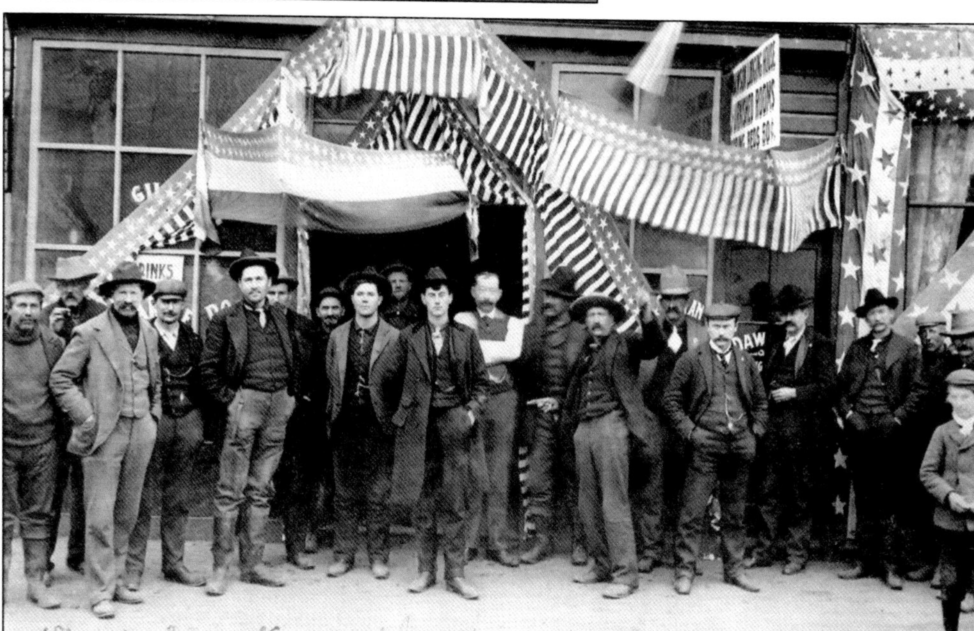

Crooks and confidence men gathered near every saloon door in Nome, as seen in this photograph taken on July 4, 1901, and law officials had their hands full keeping the rough crowds in order. Former gunslinger Wyatt Earp (fourth from right) lived in Nome from 1899 to 1901 and built the two-story Dexter Saloon with his partner Charles E. Hoxsie. (Carrie M. McLain Memorial Museum.)

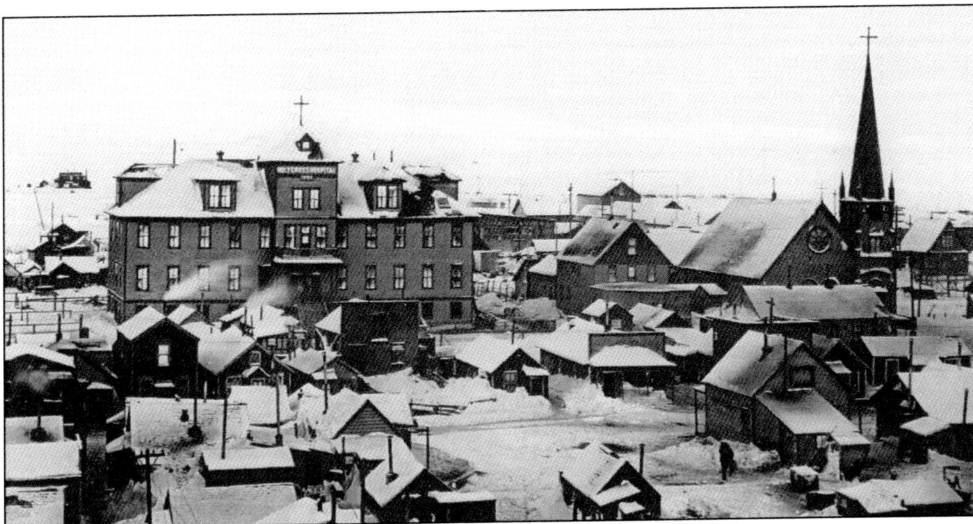

The Holy Cross Hospital (left) was the largest and most expensive building constructed in Nome in 1906. It offered modern equipment and procedures, including an x-ray machine and a sterile operating room. The Catholic-run hospital was located next door to St. Joseph Catholic Church and its parish house (right) on Fourth Avenue. (Carrie M. McLain Memorial Museum.)

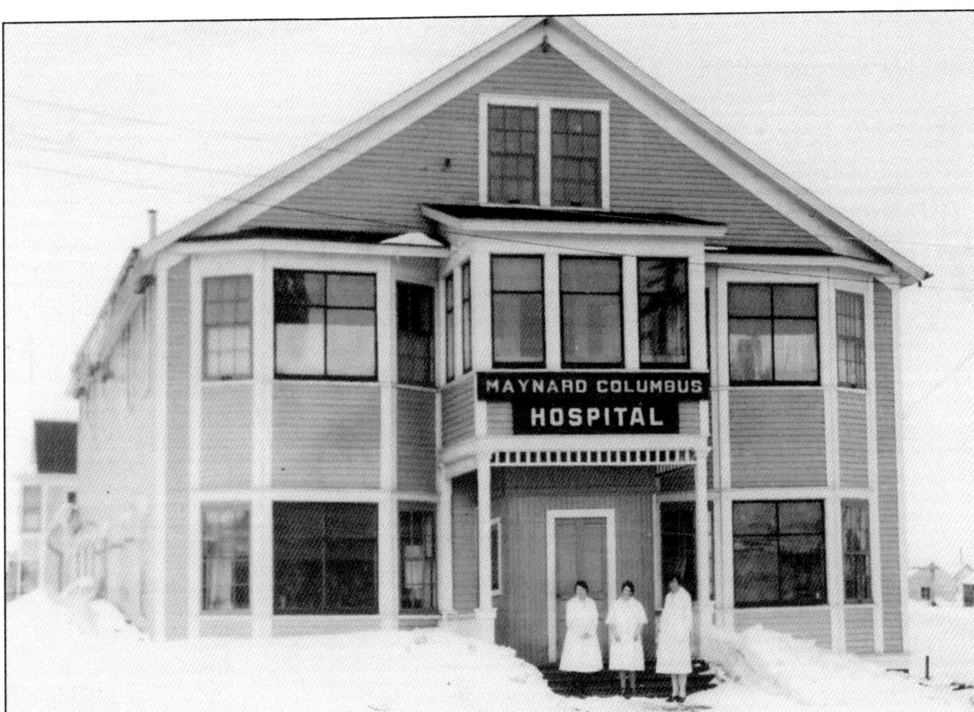

The Maynard Columbus Hospital originated as a plan in 1906 by the Women's Home Missionary Society of the Methodist Church in Columbus, Ohio, to provide medical services for Alaska Natives in Nome. Mrs. Horace Maynard provided start-up funds, and in 1927, a building from the recently closed Fort Davis was purchased for $1 and repurposed as a hospital. (Photograph by Curtis Jacobs; Carrie M. McLain Memorial Museum, Jacobs-430.)

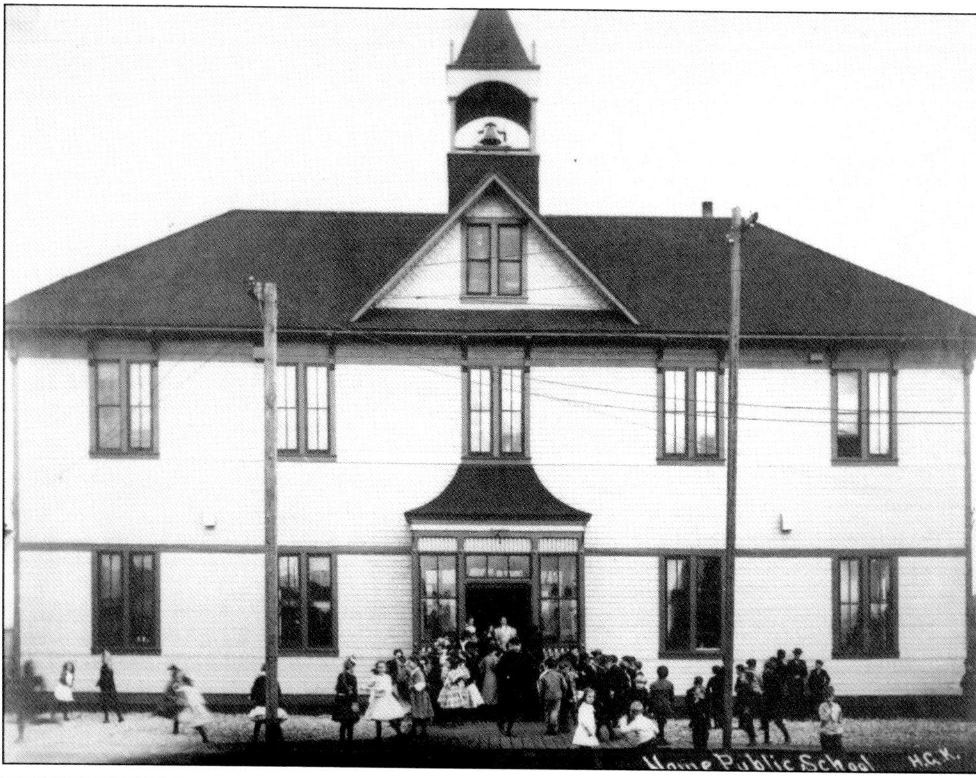

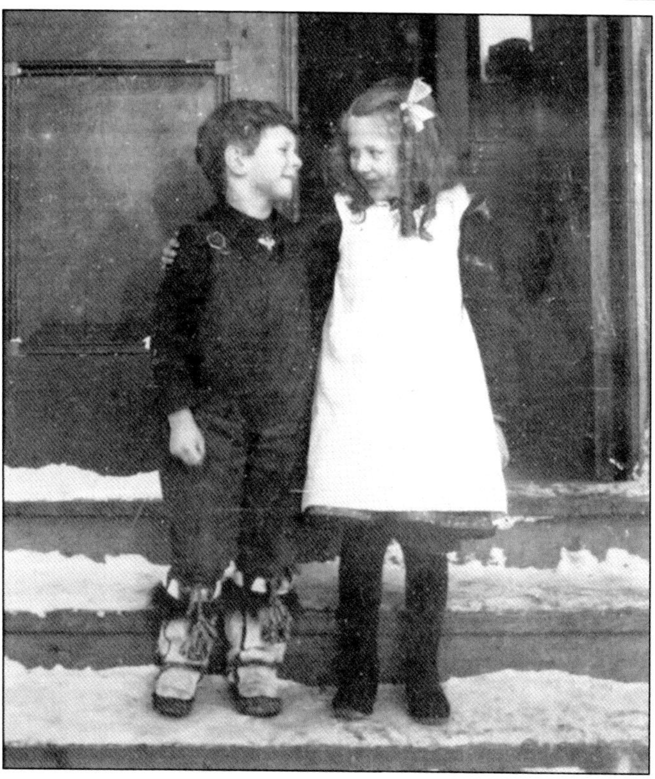

Students enjoy a break outside the Nome Public School before the bell rings. By 1907, the school featured a gym, ventilation system, and indoor lavatories for seven teachers and 350 students. (Photograph by H.G. Kaiser; Carrie M. McLain Memorial Museum, NMP-17-84.)

Jimmy Doolittle and a young friend stand on the snow-covered steps of the Nome Public School in 1904. Doolittle sold papers for the *Nome Nugget* and delivered clothes for his mother's laundry service before becoming famous as the military aviator who led the first American air attack on Tokyo during World War II. (Carrie M. McLain Memorial Museum, NMP-96-16-125.)

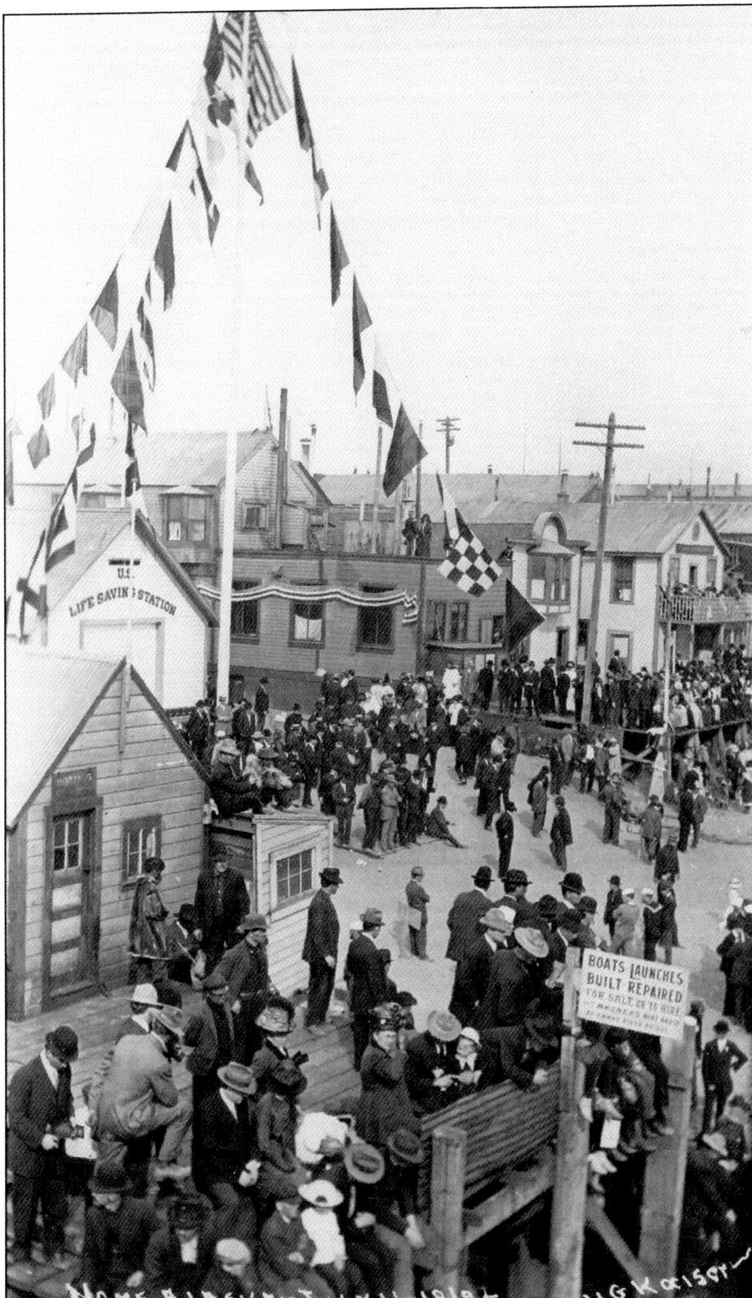

Restless crowds gather on sidewalks, hang over balconies, and perch on rooftops as they wait for the Fourth of July parade to begin in 1912. The Fourth of July marked the largest community-wide celebration of the year with parades, marching bands, foot races, donut-eating contests, skin boat races, and Alaska Native games including the blanket toss and two-foot high kick. The ban on fireworks was lifted, and the colorful devices could be set off anywhere in town except on the planked streets. Nome continues to celebrate this 100-year old-tradition every Fourth of July with a parade and street games, including a bike race, egg-and-spoon race, and a pie-eating contest. (Photograph by H.G. Kaiser; Carrie M. McLain Memorial Museum, McLain-183.)

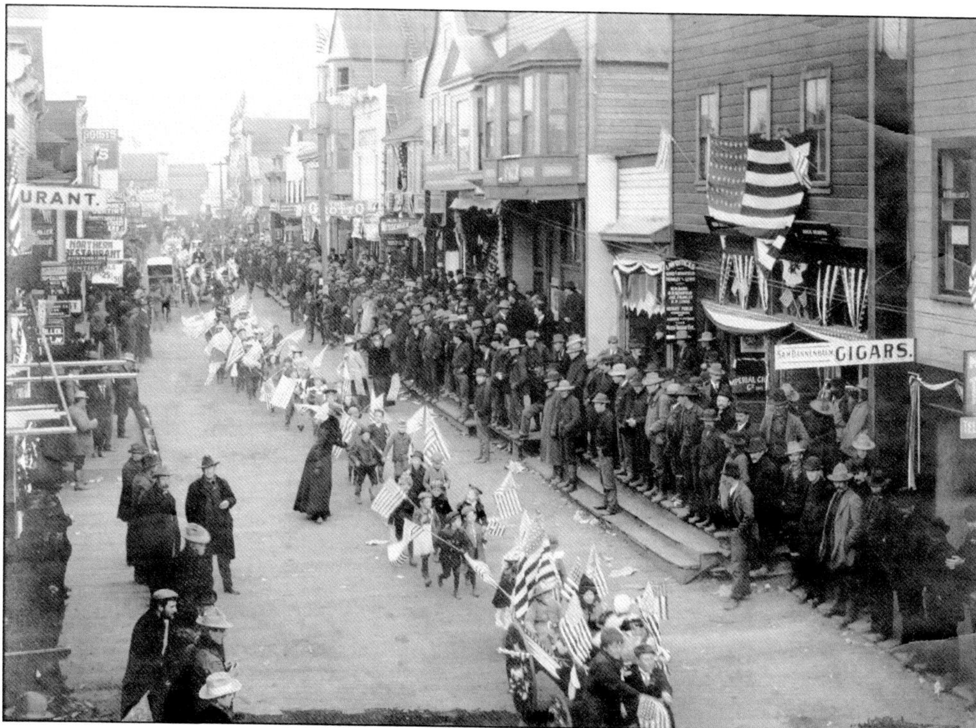

Crowds line Front Street to cheer for a parade of Nome Public School students waving flags during the Fourth of July celebration in 1901. Although Alaska would not reach statehood until 1959, that did not dampen the patriotic fervor of Nome residents, who gaily covered the town in stars and stripes. (Photograph by E.A. Hegg; Carrie M. McLain Memorial Museum, NMP-97-59-1.)

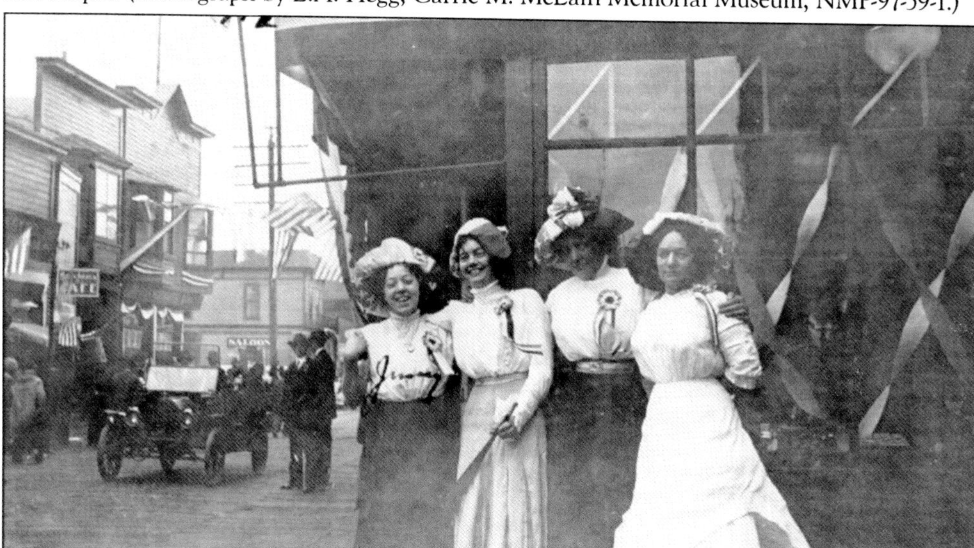

Fashionable women sporting flag hats and striped ribbons gather in front of Anvil Coffee House during the Fourth of July festivities in 1912. The "three lucky Swedes" and subsequent miners discovered rich deposits of placer gold all around the base of Anvil Mountain. This wealth inspired several storeowners to name their establishments after the propitious formation. (Carrie M. McLain Memorial Museum, NMP-11-211.)

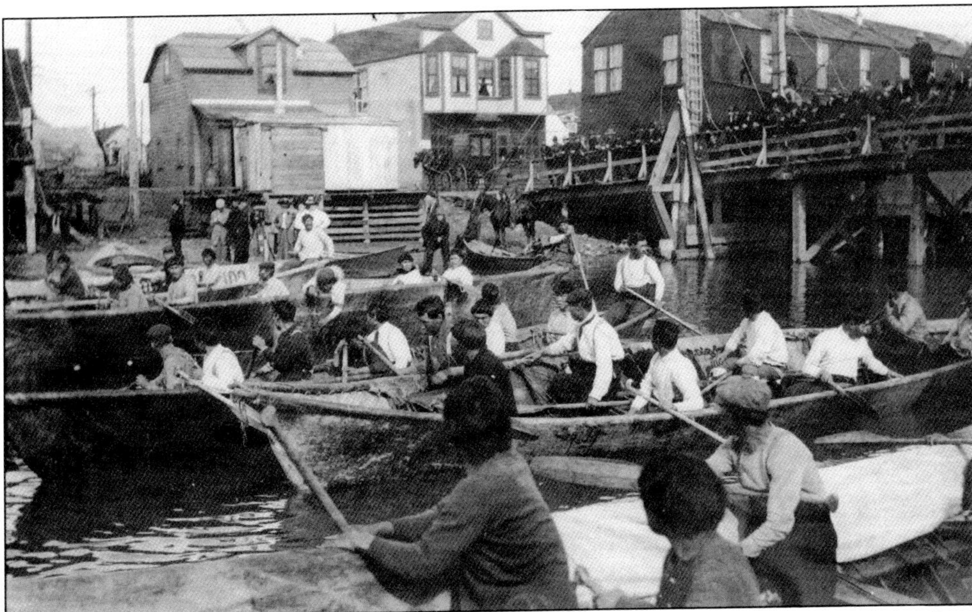

Fourth of July spectators swarm the bridge as five skin boats with eager rowers line up to race down the Snake River around 1905. Skin boats could be propelled by paddles, oars, sails, the current, or dogs on the shore attached with lines. In later years, skin boats were fitted with outboard motors, which offered greater control and decreased travel time. (Carrie M. McLain Memorial Museum, 1992.2.227.)

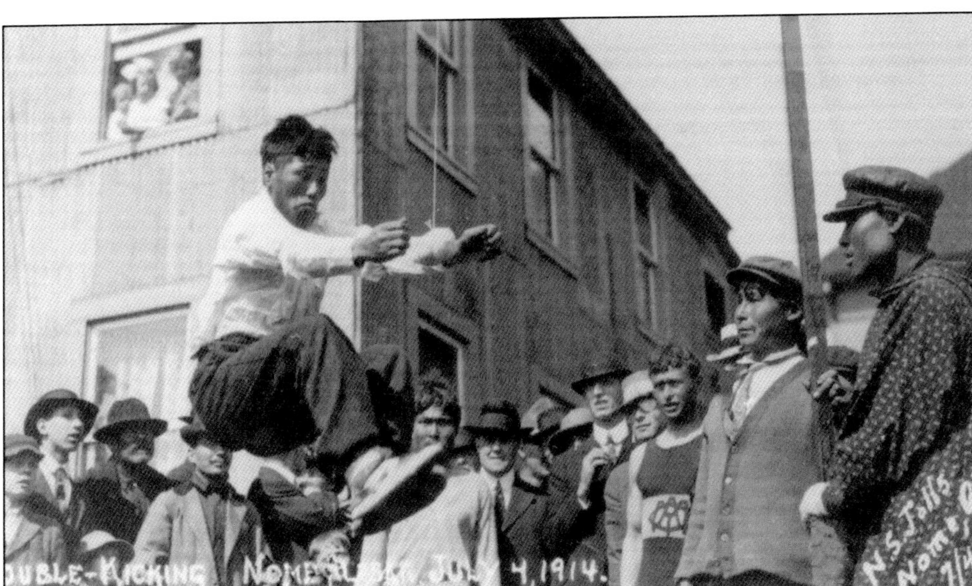

An Inupiat athlete attempts the two-foot high kick during the Fourth of July games in Nome in 1914. The two-foot high kick requires extreme balance and abdominal strength and was historically used to communicate the success of a spring hunt. Young athletes continue to participate in the two-foot high kick as well as other traditional athletic contests at the annual Native Youth Olympics. (Photograph by N.S. Jolls; Carrie M. McLain Memorial Museum, NMP-80-3-479.)

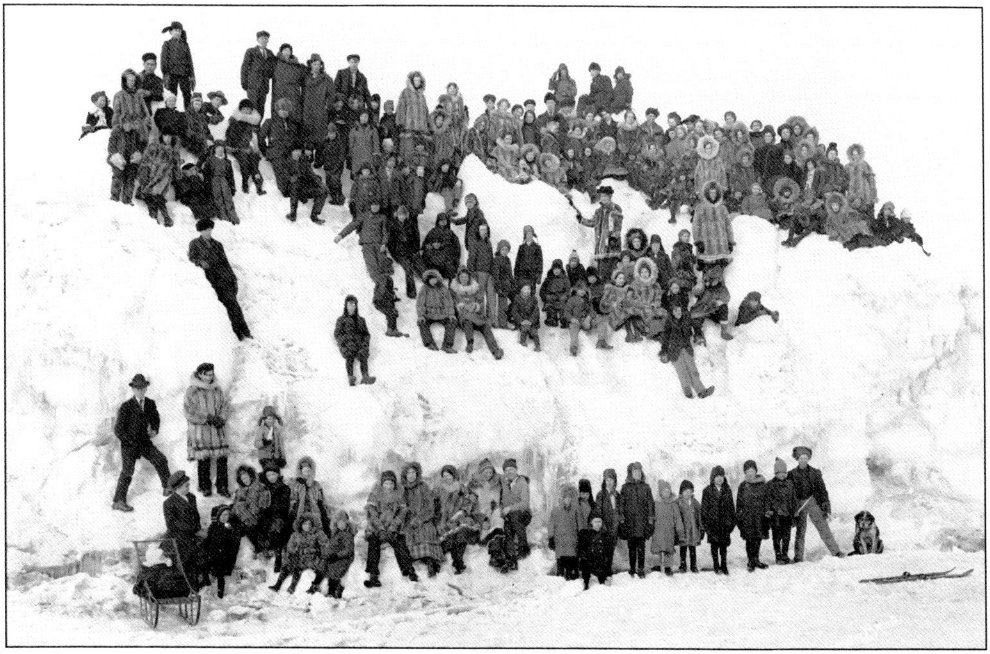

Teachers and students of the Nome Public School pose on an ice formation on the Bering Sea during a field trip on April 19, 1910. Long winters inspired creative forms of outdoor recreation including iceberg climbing and reindeer sleigh rides in addition to rounds of dances, socials, and theatrical performances. (Photograph by Lomen Brothers, Carrie M. McLain Memorial Museum, 2013.1.12.)

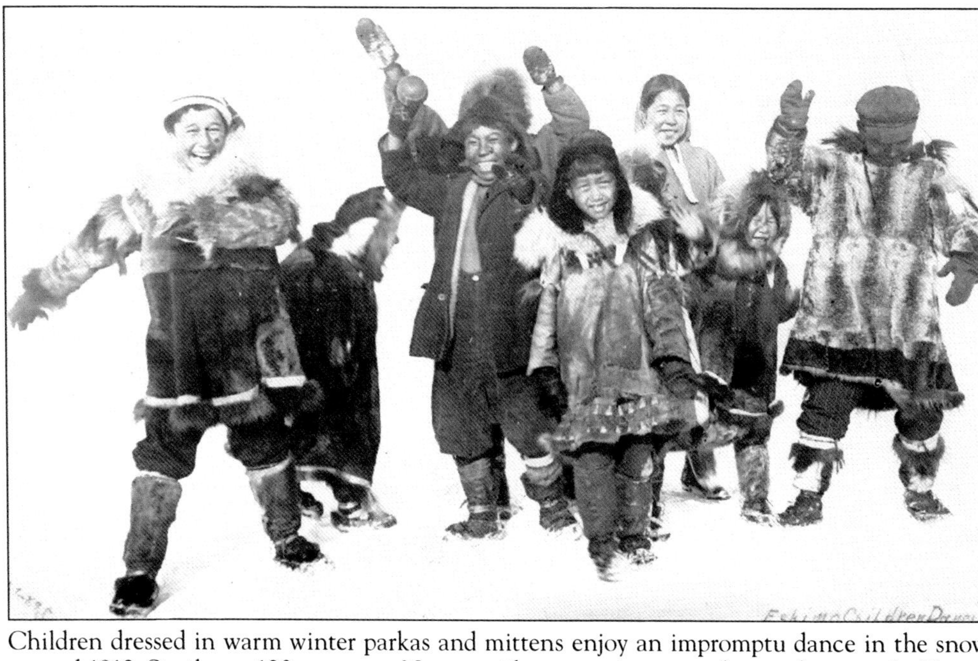

Children dressed in warm winter parkas and mittens enjoy an impromptu dance in the snow around 1910. Similar to 100 years ago, Nome residents continue to welcome the arrival of frosty winter with opportunities to cross-country ski, see the Northern Lights, and cheer on mushers in a dogsled race. (Photograph by H.G. Kaiser; Carrie M. McLain Memorial Museum, McLain-666.)

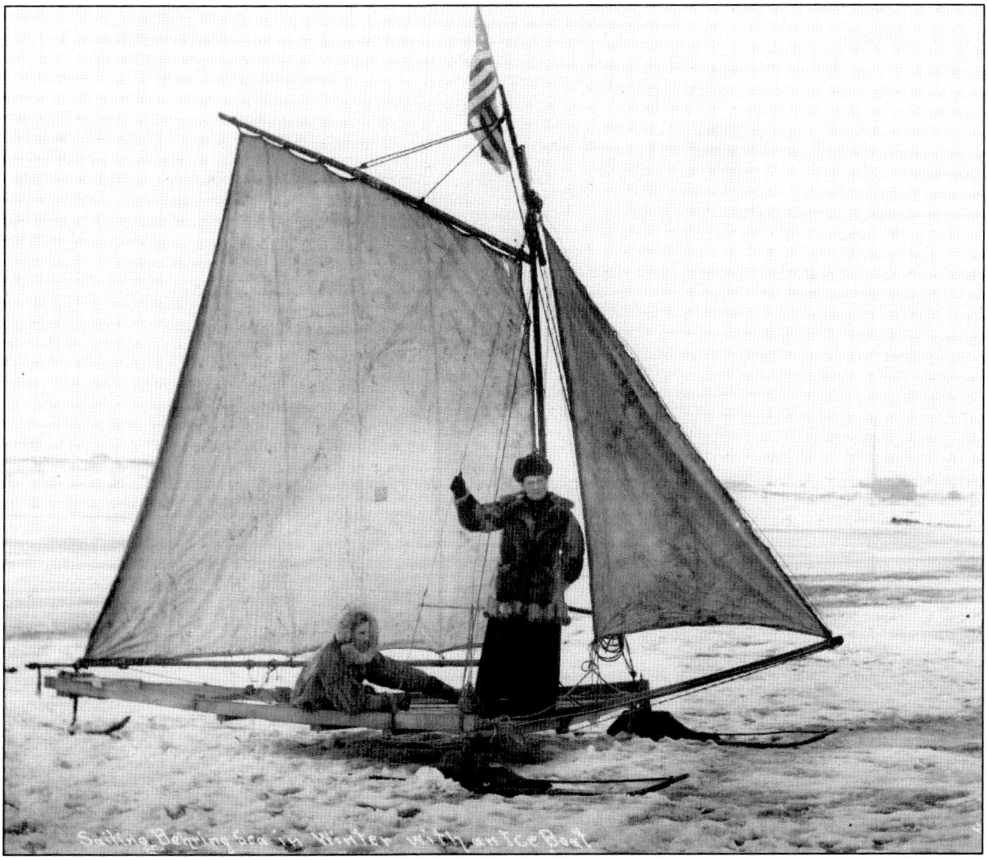

A couple attempts to hitch a ride from the wind in their ice boat equipped with sails and skis. Technological advancements in winter transportation have greatly increased over the years, and today, racers fly across the frozen sea ice in the 200-mile Nome-Golovin annual snowmachine race. (Photograph by H.G. Kaiser; Carrie M. McLain Memorial Museum, 1979.1.230.)

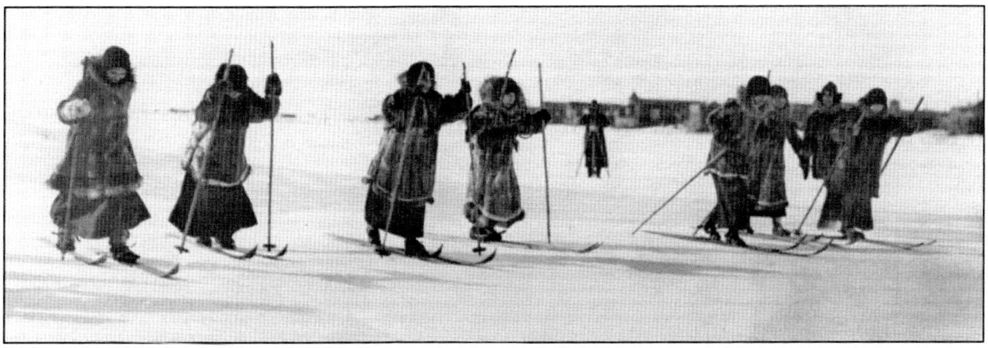

A group of Nome women take advantage of the sunshine to cross-country ski across Dry Creek around 1905. Dry Creek provided acres of clear ice for skiing and skating with a hill west of Steadman Avenue where the Sons of Norway constructed a 60-foot ski jump. (Carrie M. McLain Memorial Museum, Karrer-392.)

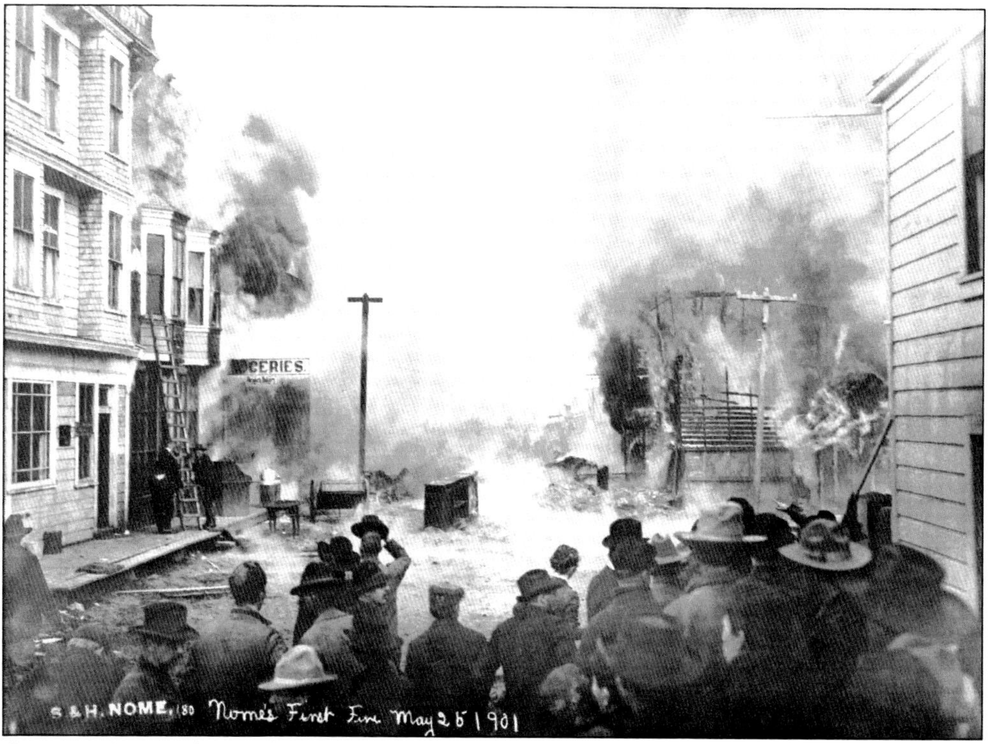

On May 25, 1901, the Golden Gate Hotel (left) triggered Nome's first major fire, which consumed eight city blocks. A bucket brigade of residents and soldiers from Fort Davis helped curb the flames, but the destruction required an emergency order of several million board feet of lumber. (Photograph by Segerstrom and Hagen; Carrie M. McLain Memorial Museum, NMP-82-5-1.)

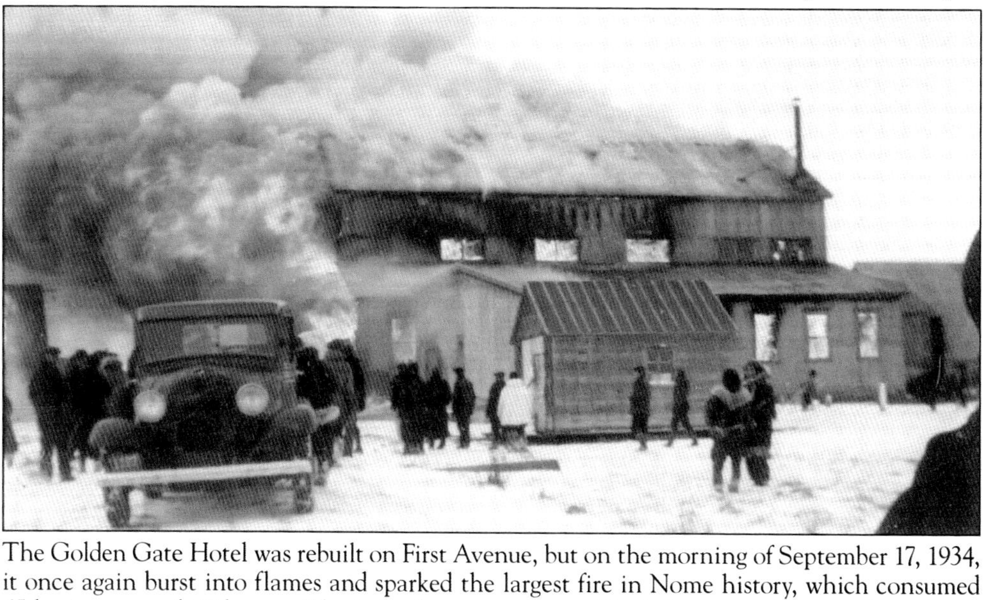

The Golden Gate Hotel was rebuilt on First Avenue, but on the morning of September 17, 1934, it once again burst into flames and sparked the largest fire in Nome history, which consumed 65 businesses and 90 homes. The federal government and Red Cross stepped in with food and building materials to aid the devastated community, which came together to rebuild the town. (Photograph by Curtis Jacobs; Carrie M. McLain Memorial Museum, Jacobs-494.)

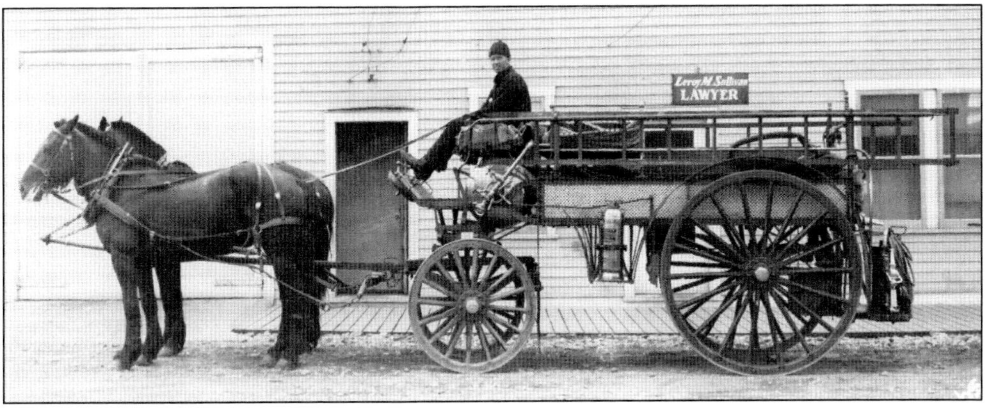

The fire department, with its horse-drawn water pumps, had its hands full attempting to keep the planked streets of downtown Nome from burning. Intake hoses could be placed in the ocean or holes kept open in the Snake River during winter. (Photograph by Curtis Jacobs; Carrie M. McLain Memorial Museum, Jacobs-82.)

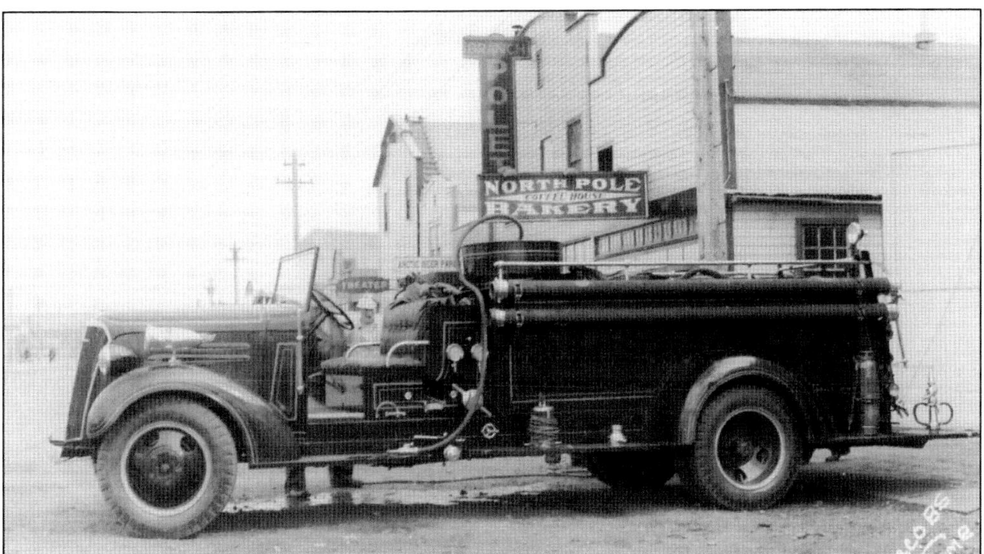

The Nome volunteer fire department acquired a modern fire truck after the disastrous fire of 1934. Following the destruction of downtown Nome, city planners decided to widen and straighten roads inside the business district to deter the spread of fires and increase accessibility for automobiles. (Photograph by Curtis Jacobs; Carrie M. McLain Memorial Museum.)

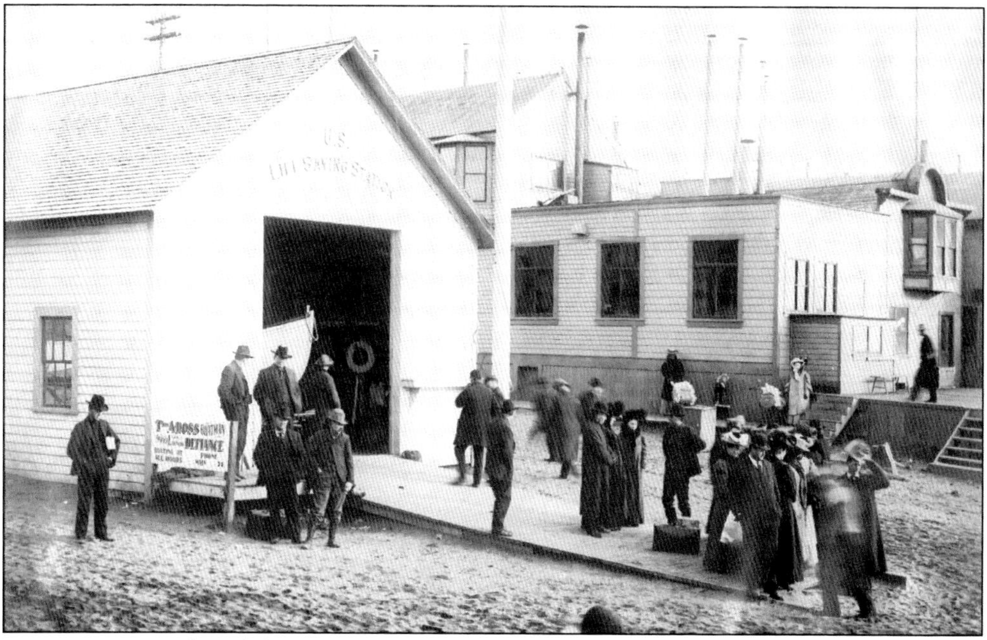

Passengers wait to board the SS *President* outside the US Life Saving Station in Nome on October 14, 1907. The station was established in 1905 under the leadership of Thomas Ross and his crew of seven Scandinavian-born seamen. Lifeboats were mounted onto wagon frames and pulled to the beach by the lifesaving crew, who jumped inside once they launched the boats. The lifesaving crew assisted in ferrying people from ships to the beach, lent aid to firefighters, towed grounded vessels, and helped search for missing aircraft. (Photograph by Otto D. Goetze; Carrie M. McLain Memorial Museum, 1999.12.1.)

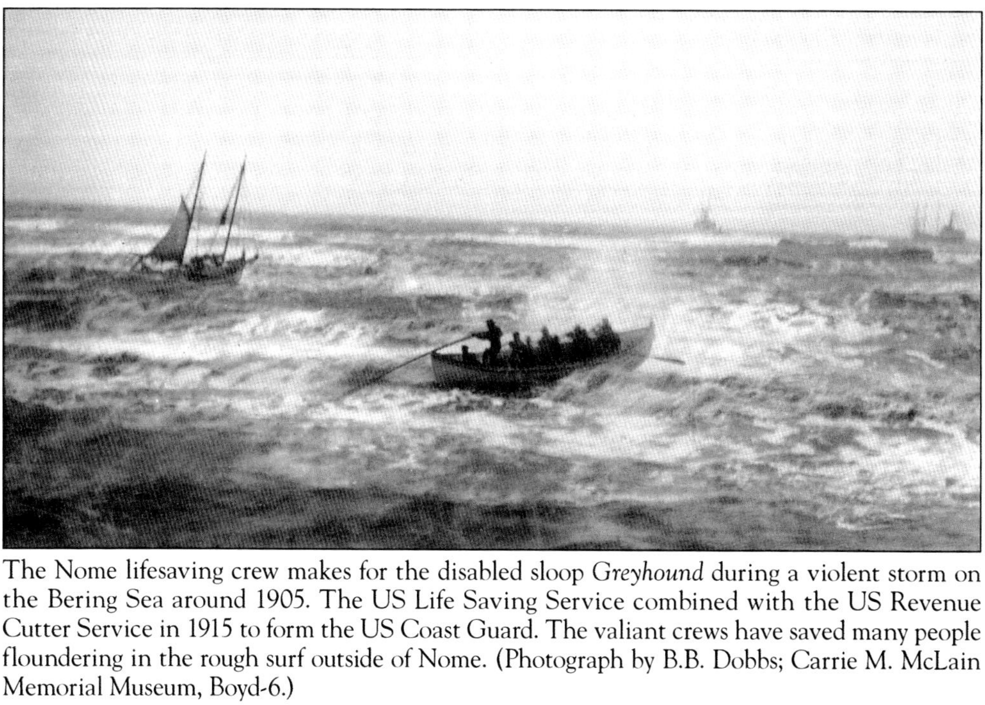

The Nome lifesaving crew makes for the disabled sloop *Greyhound* during a violent storm on the Bering Sea around 1905. The US Life Saving Service combined with the US Revenue Cutter Service in 1915 to form the US Coast Guard. The valiant crews have saved many people floundering in the rough surf outside of Nome. (Photograph by B.B. Dobbs; Carrie M. McLain Memorial Museum, Boyd-6.)

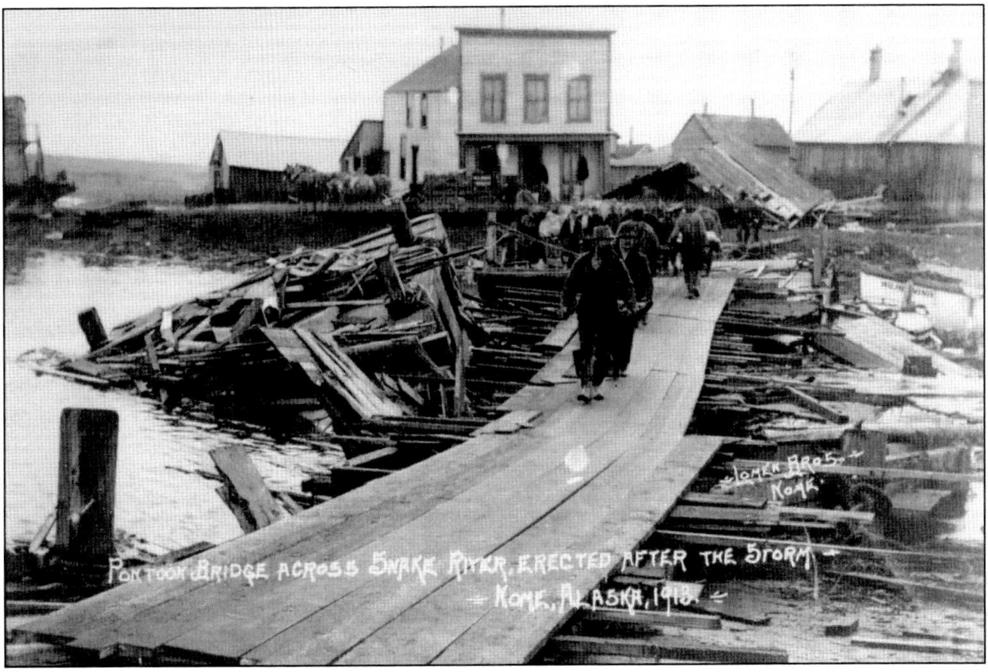

Nome residents cross the Snake River on a temporary pontoon bridge made from salvaged lumber following the Bering Sea storm of October 1913. The storm raged along the Nome coastline and tore buildings off foundations, smashed houses, and hammered 10 blocks of buildings into matchwood along Front Street. (Photograph by Lomen Brothers; Carrie M. McLain Memorial Museum, NMP-19-96.)

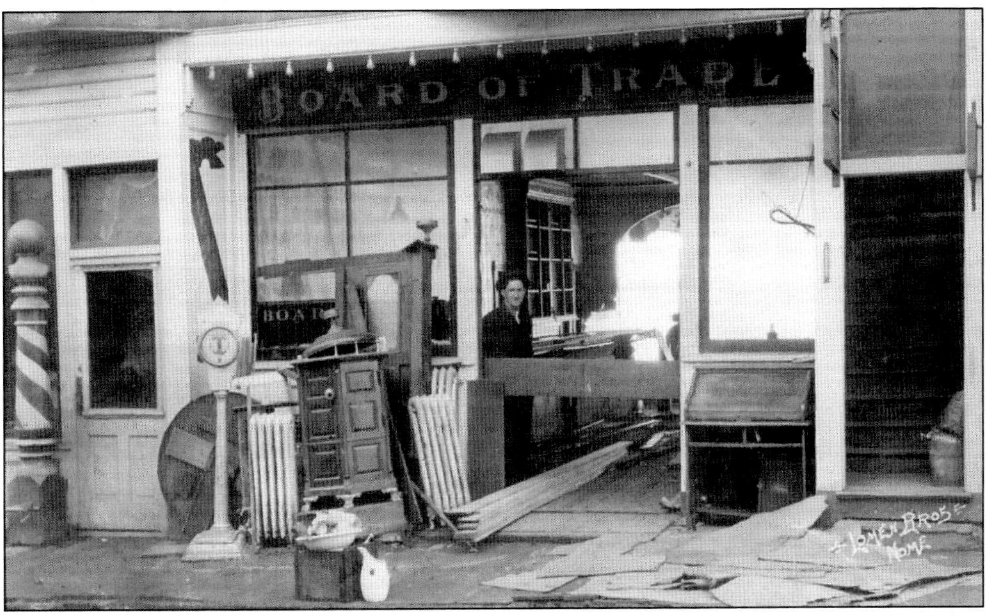

The Board of Trade saloon on Front Street suffered heavy damage during the storm of October 1913. The longest operating saloon in Nome has been rebuilt several times since it opened in 1900, and has retained its popularity with visitors and locals alike. (Photograph by Lomen Brothers; Carrie M. McLain Memorial Museum, Rolando-83.)

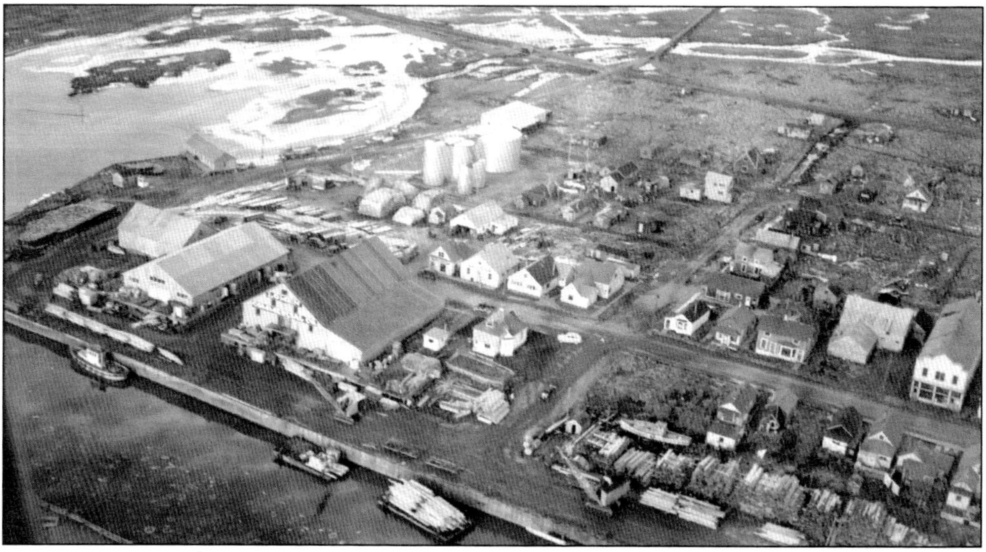

Lomen Commercial Company warehouses and lighters transporting lumber are shown here on the east side of the Snake River around 1930. G.J. Lomen and his son Carl arrived in Nome in 1900 and were followed soon after by Carl's mother, Julie Joys Lomen; sister Helen; and brothers Ralph, Harry, George, and Alfred. The entrepreneurial family became involved in many commercial enterprises in Nome including law, photography, lighterage, retail, and reindeer. (Carrie M. McLain Memorial Museum.)

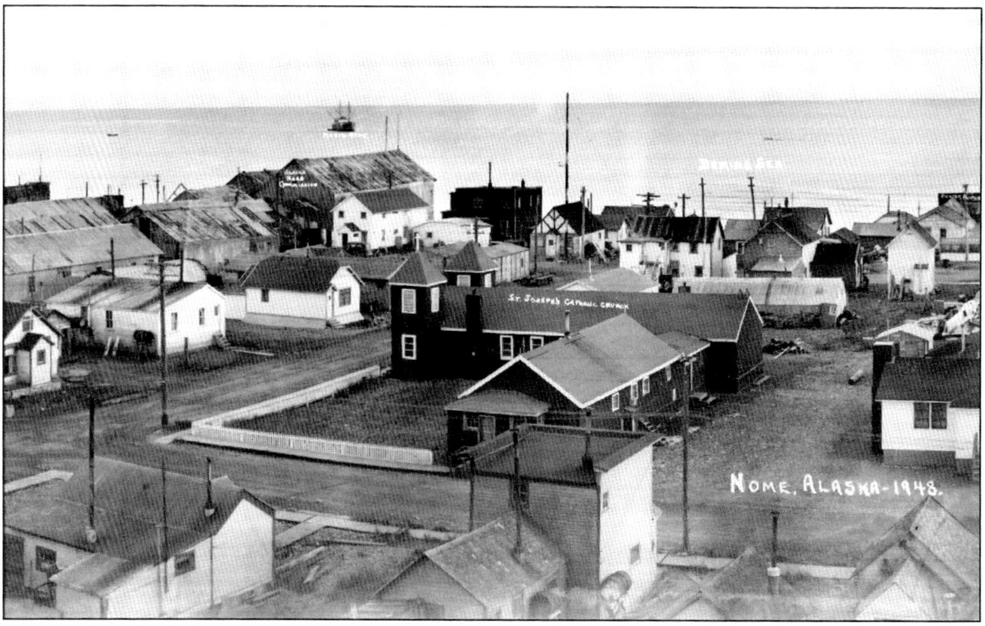

A view down Steadman Street in Nome in 1948 shows the new St. Joseph Catholic Church and parish house, which were built in a modern style with simple architectural details. The tall building of the Alaska Road Commission is on the left, and the *North Star* supply ship is anchored behind in the Nome roadstead. Metal Quonset huts with curved roofs and corrugated siding have been relocated from Marks Air Force Base and appear interspersed between the buildings. (Carrie M. McLain Memorial Museum.)

Six

Nome Merchants

Enterprising merchants arrived in Nome alongside gold prospectors eager to stake a share in the wealth of the flourishing town. Over 120 stores and businesses lined the two main thoroughfares of Front Street and Steadman Avenue. Two-story buildings with false-front facades jostled for customers' attention with offers of clothing and lodging, building supplies and fresh produce, and everything needed to outfit a miner, furnish a home, or alleviate failed dreams of striking it rich.

The clink of glasses, shuffle of cards, and spin of a roulette wheel could be heard inside more than 50 saloons in gold rush Nome. Drinking establishments along Front Street were some of the largest and finest buildings in town, with mahogany bar fixtures, stained-glass windows, and crystal chandeliers. Hotels promised modern amenities and fine-dining experiences for discerning patrons. In 1905, the Golden Gate Hotel featured a menu with lobster salad, spring lamb, and orange sherbet set to the melody of a live orchestra.

Capitalists enjoyed good returns on their investments in the three principal banks of early Nome. The Alaska Banking and Safe Deposit Company, Bank of Cape Nome, and Miners and Merchants Bank of Nome all purchased gold, rented safety deposit boxes, and examined ore for precious metals in their assay offices. The Nome bar included a healthy number of lawyers whose offices represented large mining corporations and handled suits over mining claims, patents, and water rights along the Nome River.

Professional photographers, including Frank H. Nowell and Otto D. Goetze, discovered creative inspiration in the bustle of activities along the streets and beaches. Photographers opened multiuse stores with portrait studios, darkrooms, and display windows to sell prints and postcards to eager tourists. Reporters worked the crowds to get the latest information on new gold strikes and any word from the outside. Over 10 newspapers operated during the first five years of Nome, with a troop of paperboys who peddled weekly and semiweekly editions that documented the ever changing community.

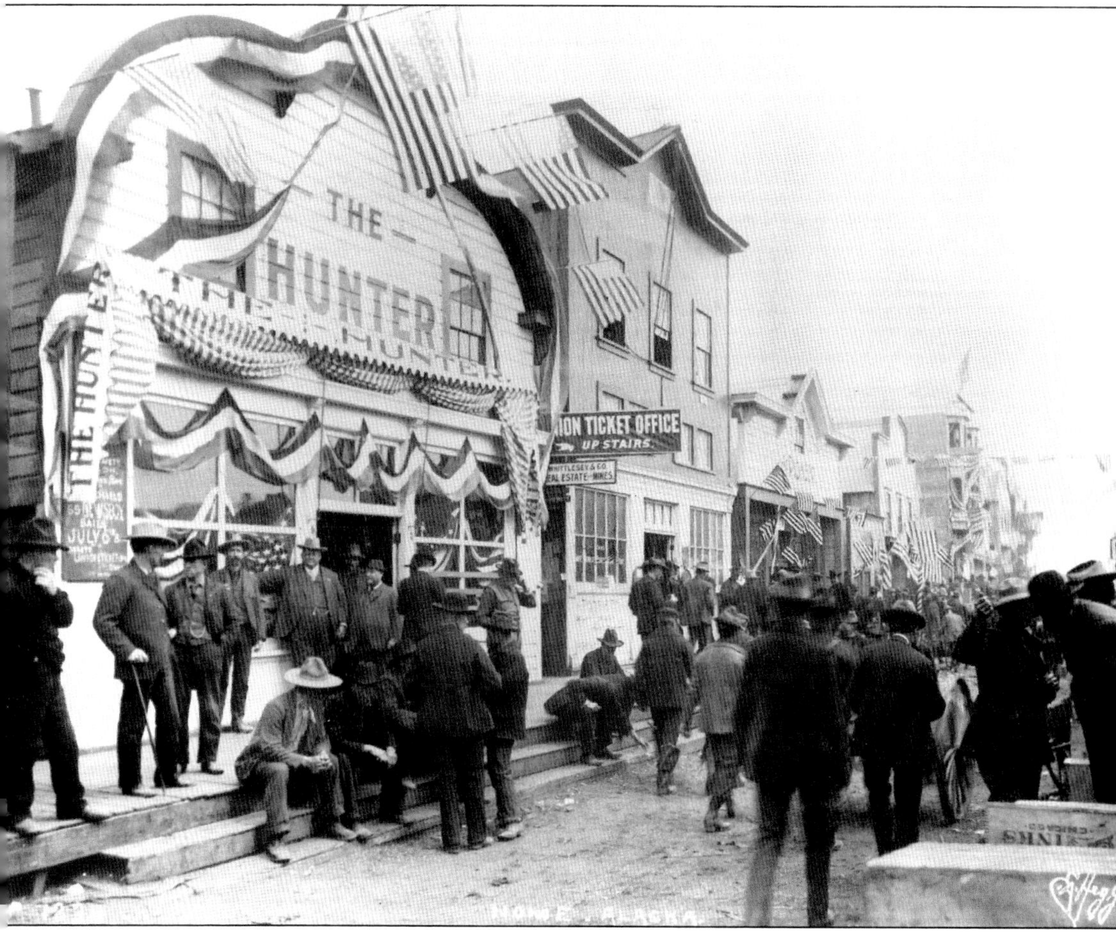

This photograph shows crowds of spectators in front of stores dressed with colorful flags and bunting during Nome's Fourth of July festivities in 1900 or 1901. In 1905, a fire raced along the wooden sidewalks of Front Street and consumed almost 50 businesses in the center of town. Resilient business owner Leonard F. Woods re-raised the rafters of the Hunter Saloon, shown here at the corner of Front Street and Hunter Way, less than a month after it was consumed by flames. (Photograph by E.A. Hegg; Carrie M. McLain Memorial Museum, NMP-1334.)

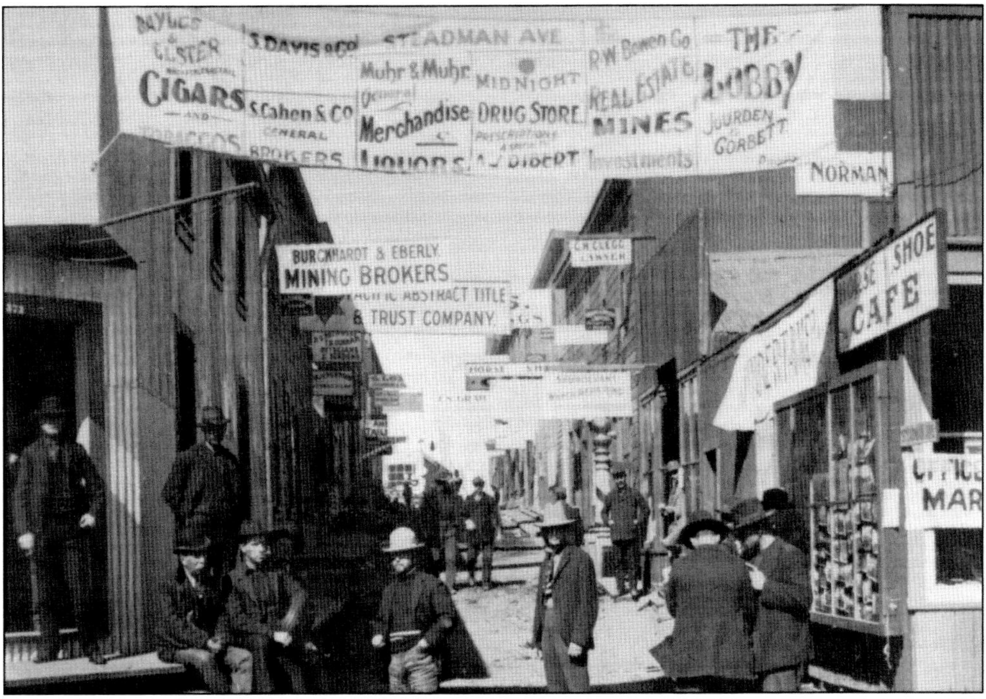

The booming city of gold rush Nome attracted a wide variety of shop owners and business professionals who were quick to hang out their signboards, as seen here in July 1900. The vibrant business district boasted six bakers, five laundries, two tin shops, four watchmakers, two sign painters, eleven physicians, sixteen lawyers, four bathhouses, one confectionery store, and one massage artiste. (Carrie M. McLain Memorial Museum, NMP-82-61-53.)

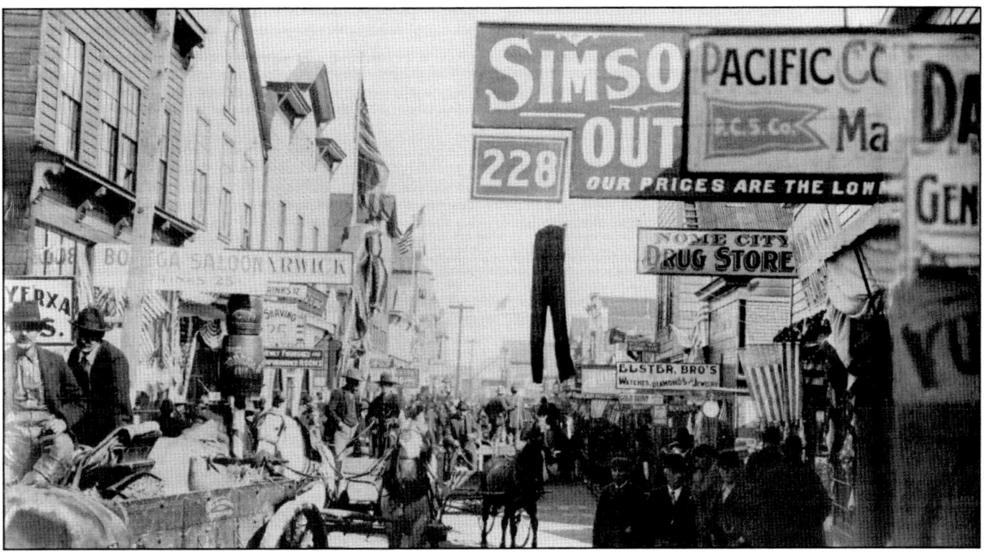

Two-story clapboard buildings leaned over the main streets of Nome. Upstairs floors were regularly rented as living quarters to lawyers, doctors, dentists, and other businessmen. Merchants at street level vied for space and potential customers with offers of clothing, pharmaceuticals, meals, lodging, silent film showings, stiff drinks, and steamship tickets back home. (Carrie M. McLain Memorial Museum, NMP-17-71.)

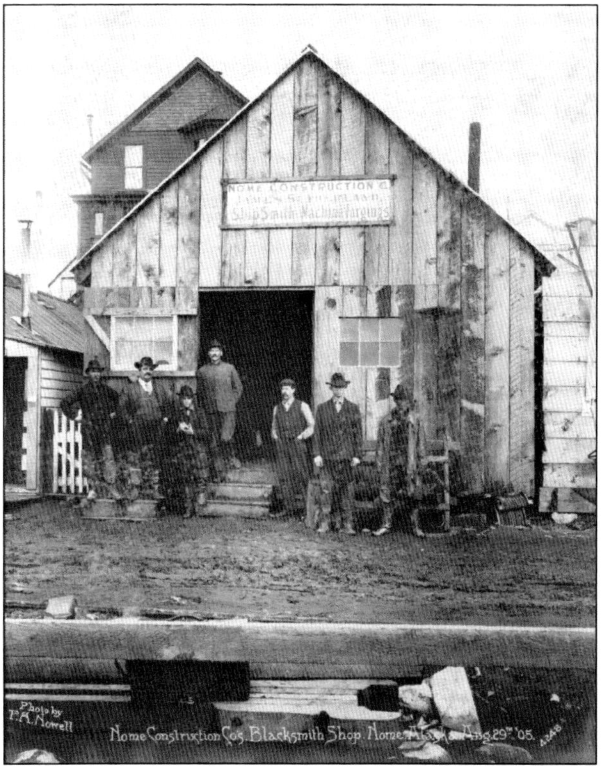

Blacksmiths at the Nome Construction Company, run by James Sutherland, forged a variety of metal parts for ships and mining apparatus, as seen here in 1905. The first businesses in Nome operated out of tents and thrown-together structures. More successful ventures were eventually housed in buildings with decorative storefronts and plate-glass windows downtown. (Photograph by F.H. Nowell; Carrie M. McLain Memorial Museum, NMP-2379.)

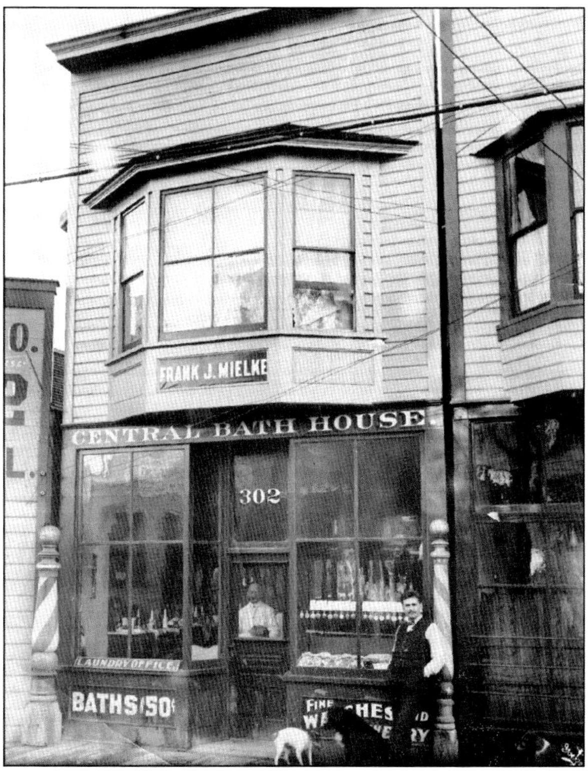

Frank Mielke ran the profitable Central Bath House, which also offered barbershop services and a selection of fine jewelry and watches, as seen here in 1905. Frank Mielke and his wife, Lizzie, relocated from Seattle to Nome in 1900 and quickly became prominent in the community, with Frank serving as a city councilman, store owner, and property manager. Their adventurous daughter Clara was born in 1903, and in 1979, she bequeathed her family's collection of artifacts and photographs to the Carrie M. McLain Memorial Museum. (Photograph by Otto D. Goetze; Carrie M. McLain Memorial Museum, 96.24.213.)

Mercantile stores, such as the Beau Mercantile Company shown here between 1901 and 1905, offered a wide range of building supplies and dry goods from pot-bellied stoves, pans, and watering cans to rifles, rakes, and handfuls of nails. John L. Beau moved from Wisconsin to Alaska where he first sold shoestrings at St. Michael before he moved to Nome and prospered as a miner, fur trader, and storeowner. (Carrie M. McLain Memorial Museum, NMP-17-46.)

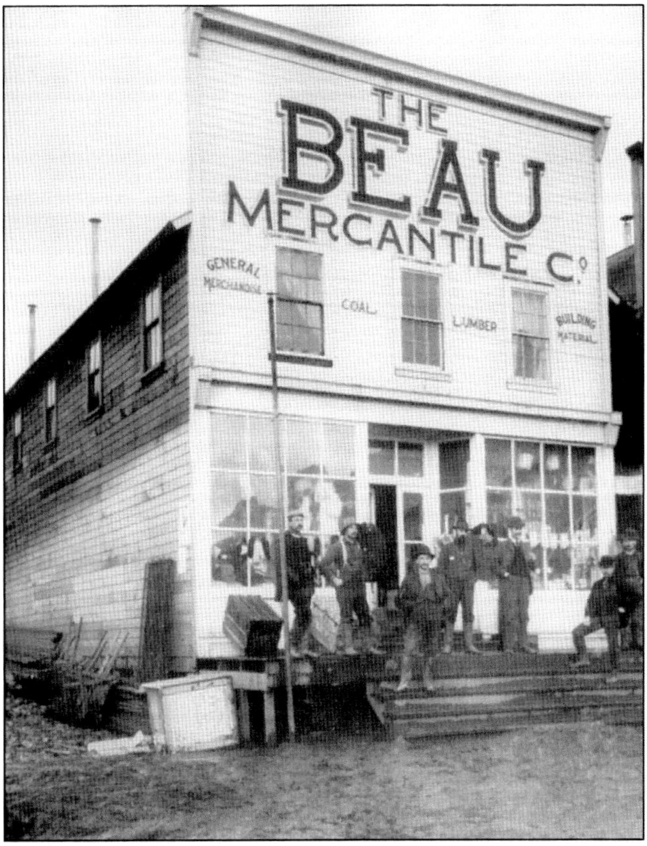

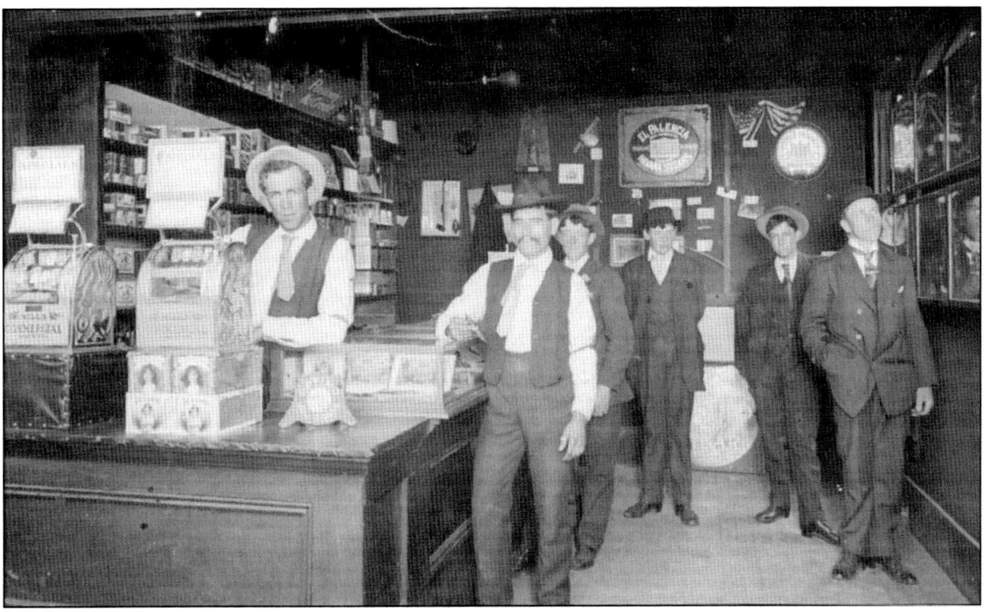

Customers are dressed in their finest for a visit to the cigar store around 1904. The countertop holds two poker card flipping machines that awarded cigars for winning hands of cards. (Carrie M. McLain Memorial Museum, NMP-80-3-170.)

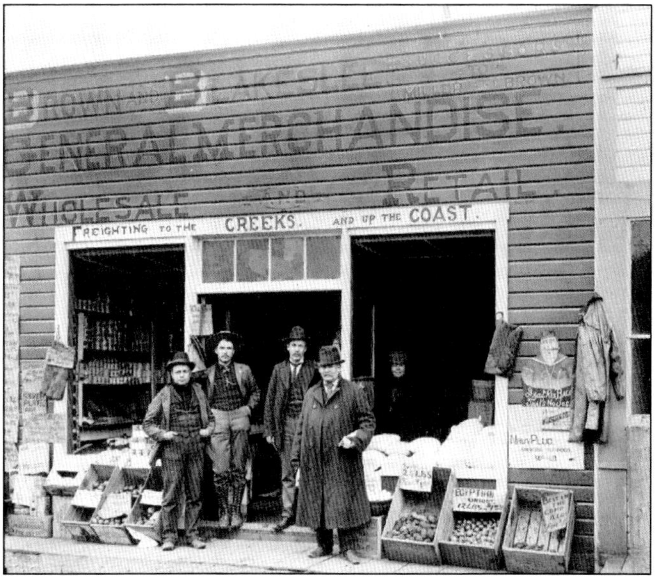

The general merchandise store of Brown and Blakeslee sold everything a prospector might need for his camp kitchen, including silver plates, cream cheese, syrup, salted bacon, canned beans, and 20 pounds of potatoes for $1. Miners who ran out of supplies on the tundra could stop at one of the many roadhouses across the Seward Peninsula, which offered hot meals, cots, and often a barn with straw for their horses. (Carrie M. McLain Memorial Museum, McLain-323.)

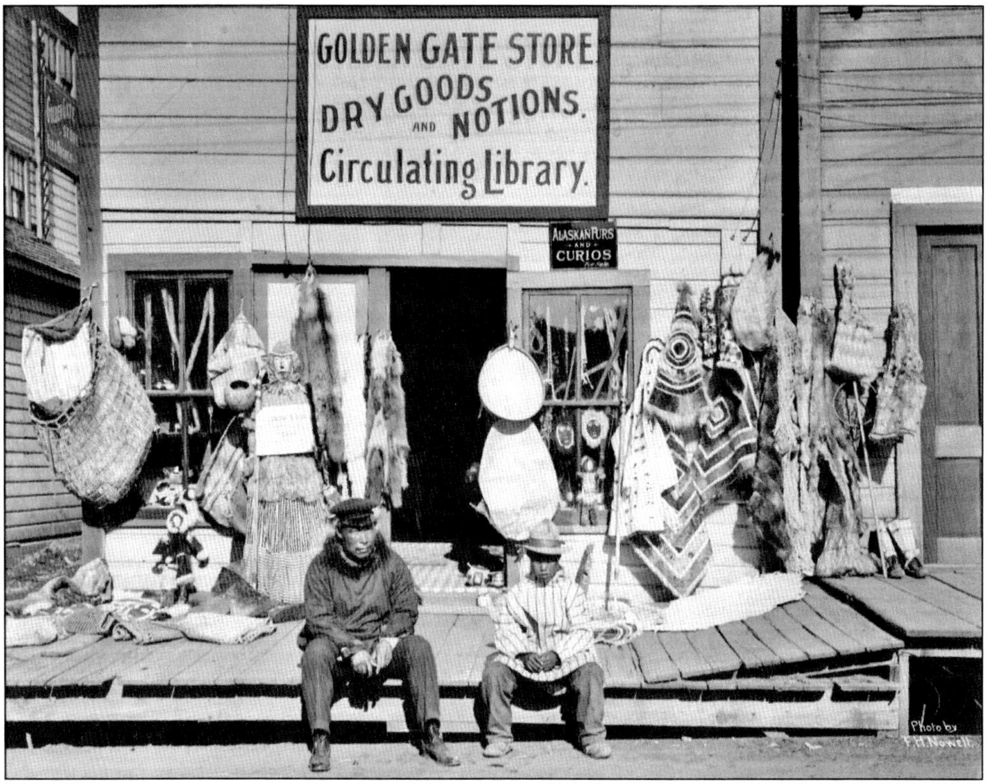

The Golden Gate Store run by Ella Ongmen offered a wide variety of Alaska Native items for sale, including gut-skin parkas, woven grass bags, drums, masks, dolls, and decorative fur wall hangings, as displayed here in September 1906. Visitors to Nome have continued to seek out Alaska Native artwork at historic stores over the years, including the A. Polet store, Arctic Trading Post, Chukotka-Alaska, and Maruskiya's. (Photograph by F.H. Nowell, Nome; Carrie M. McLain Memorial Museum, NMP-2403.)

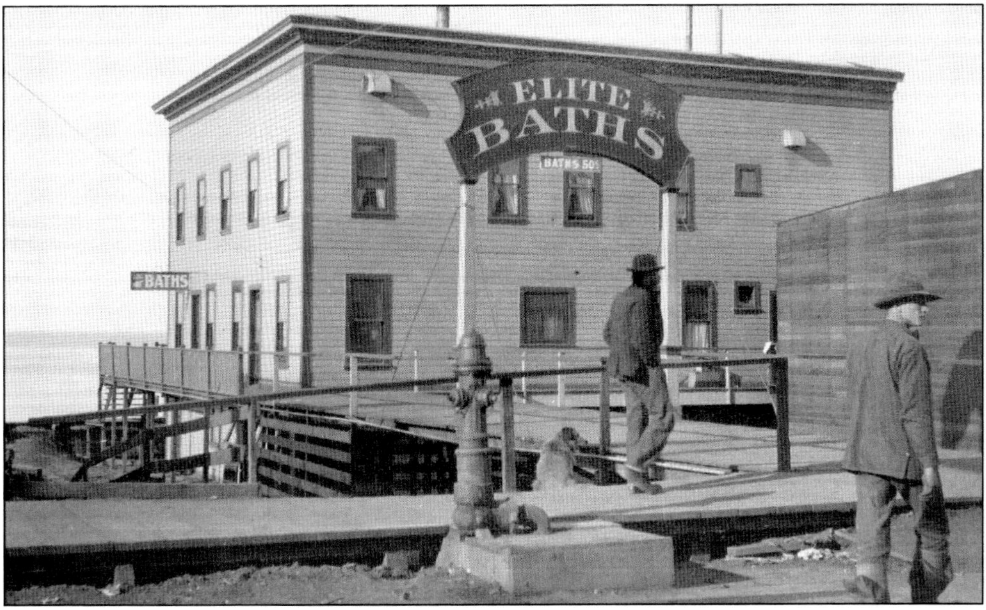

In 1906, dust-covered patrons at Elite Baths in Nome could soak in a tub for 50¢ with drains that conveniently flushed directly into Norton Sound. J.T. Thornquist operated Elite Baths and sang second bass in the Elite Quartet. Elite Baths was one of the first shorefront businesses to be destroyed in the disastrous Bering Sea storm of October 1913. (Carrie M. McLain Memorial Museum, 2016.6.)

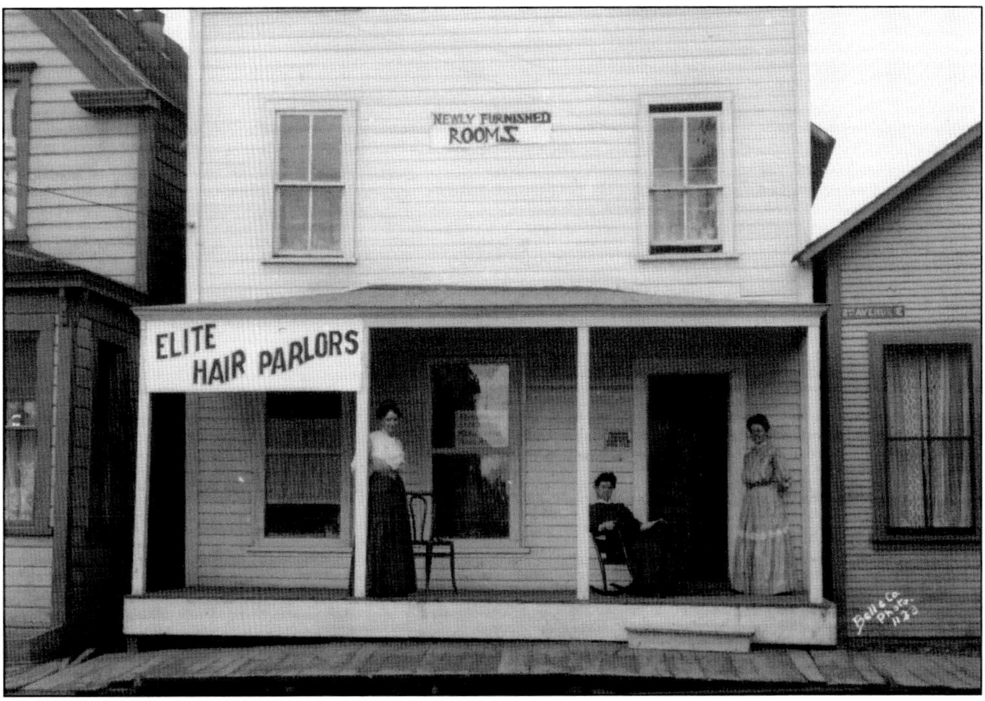

Nome women in search of a fashionable hairstyle, manicure, or facial massage could visit Elite Hair Parlors on Second Avenue, shown here around 1903–1909. Less discerning patrons could visit the beach, where barbers enjoyed brisk business with a chair and straight razor. (Photograph by Arthur L. Bell; Carrie M. McLain Memorial Museum, Bell-N-373.)

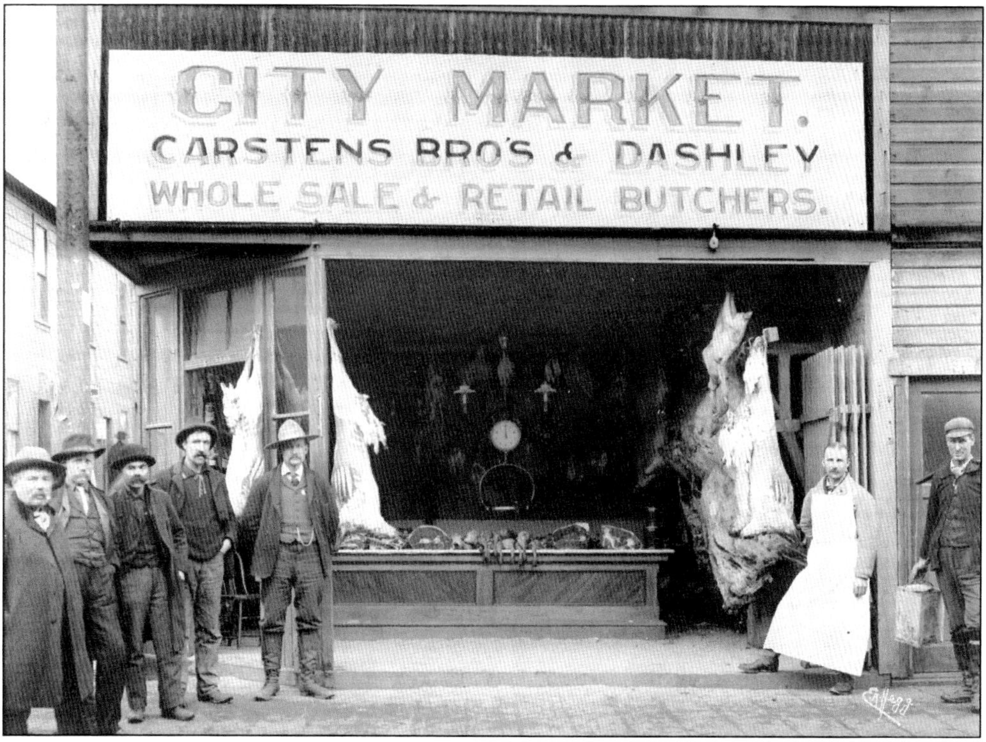

Customers could choose from birds, beef, and miscellaneous meats for sale at the City Market in Nome in 1900–1901. Butchers purchased hundreds of live cattle, sheep, and hogs from incoming steamships and stockpiled the meat for the seven isolated months of winter. (Photograph by E.A. Hegg; Carrie M. McLain Memorial Museum.)

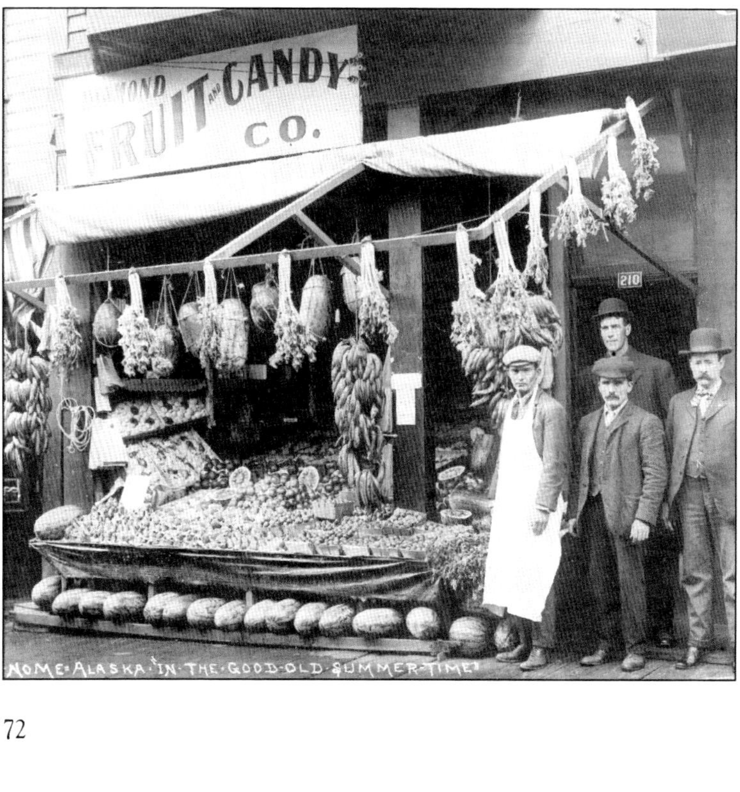

Frequent arrival of steamships during the summer supplied Nome grocers, such as the Diamond Fruit and Candy Company, with packed crates of fresh fruits, vegetables, and dairy products. Residents carefully preserved the prized produce in hopes the food would last and provide a safeguard against scurvy during the long winter season. (Photograph by Huey; Carrie M. McLain Memorial Museum, 2013.1.)

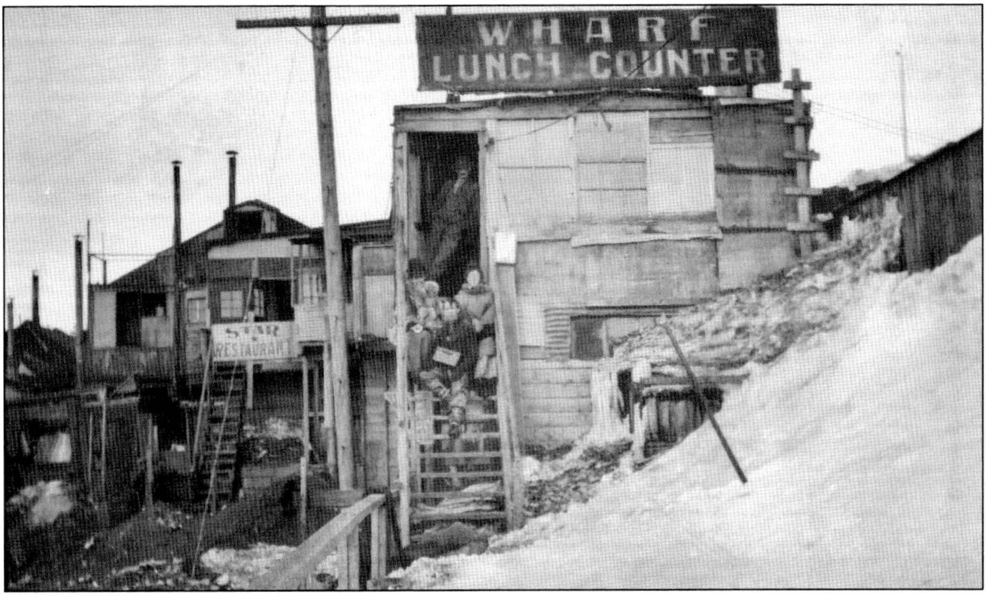

The high cost of shipping lumber to Nome prompted merchants strapped for cash to piece together buildings from scraps of metal and wood. Proprietors of the Wharf Lunch Counter and the Star Restaurant, seen here in 1909, assembled their establishments from wood planks, corrugated metal, and various sized windows. (Photograph by Arthur L. Bell; Carrie M. McLain Memorial Museum, Bell-N-369.)

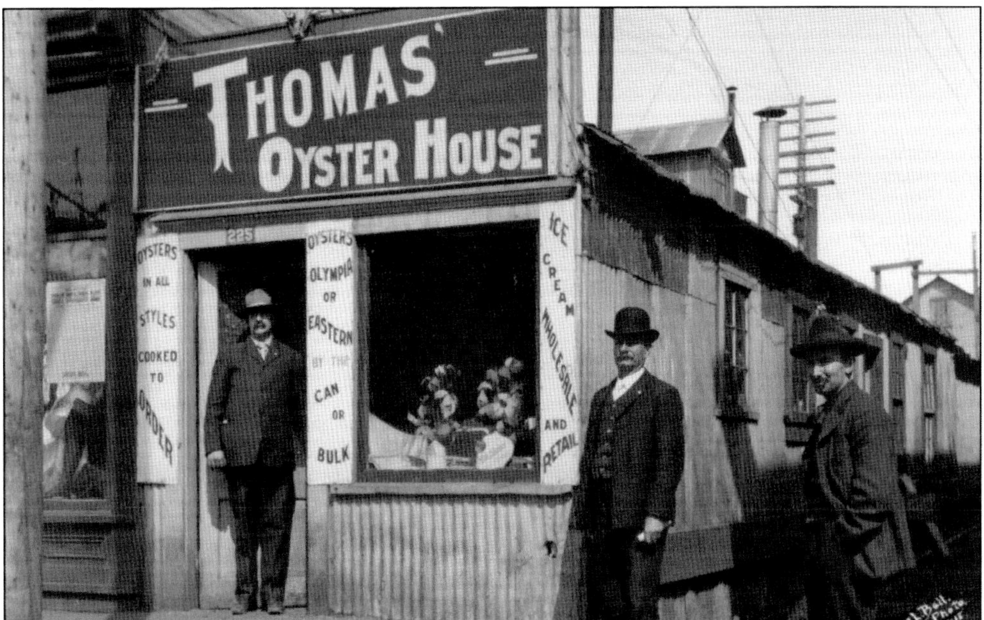

An oyster craze swept across North America during the late 1880s, and oyster bars became popular gathering places in towns across the nation. Thomas' Oyster House in Nome offered fresh or canned oysters from the Pacific and Atlantic coasts. Before checking out, customers could choose from 18 flavors of milk shakes to complete their meal. (Photograph by Arthur L. Bell; Carrie M. McLain Memorial Museum, Bell-N-366.)

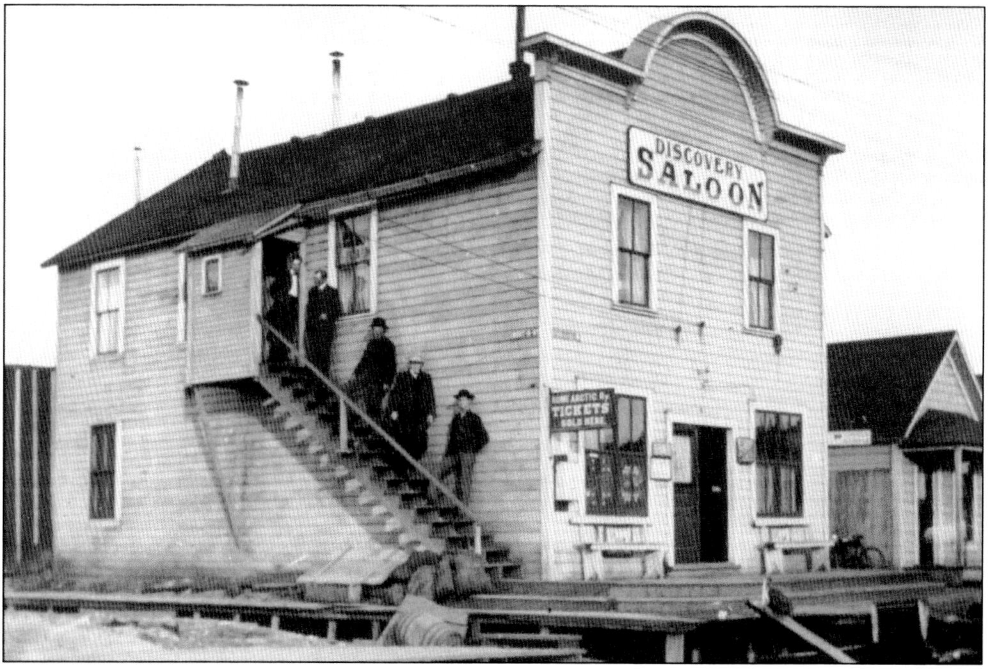

The Discovery Saloon was built in September 1899 by Jafet Lindeberg, one of the "three lucky Swedes," who discovered gold and helped found Nome. The saloon was known as a family-friendly establishment, the only saloon in Nome where a woman other than "ladies of the evening" could belly up to the bar. The Discovery Saloon became a private residence around 1915 and has since been restored by its current owners to its original false-front grandeur. (Carrie M. McLain Memorial Museum, NMP-66-54-405.)

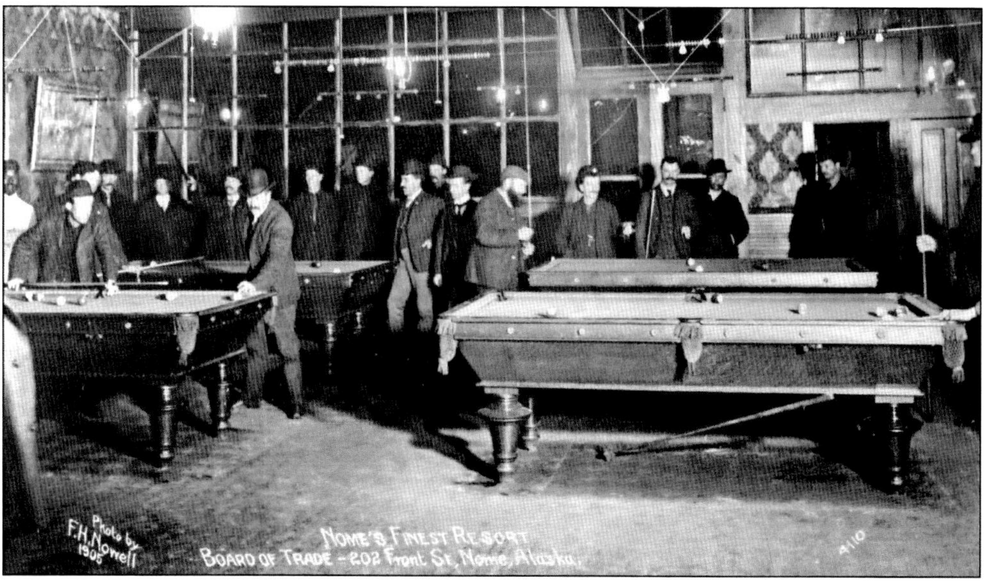

Men place bets on pool games during a busy evening in the electrically lit pool hall of the Board of Trade saloon in 1905. Saloons offered various enticements to those flush with gold, including drinks, meals, companionship, and betting games. (Photograph by F.H. Nowell; Carrie M. McLain Memorial Museum, NMP-12-2.)

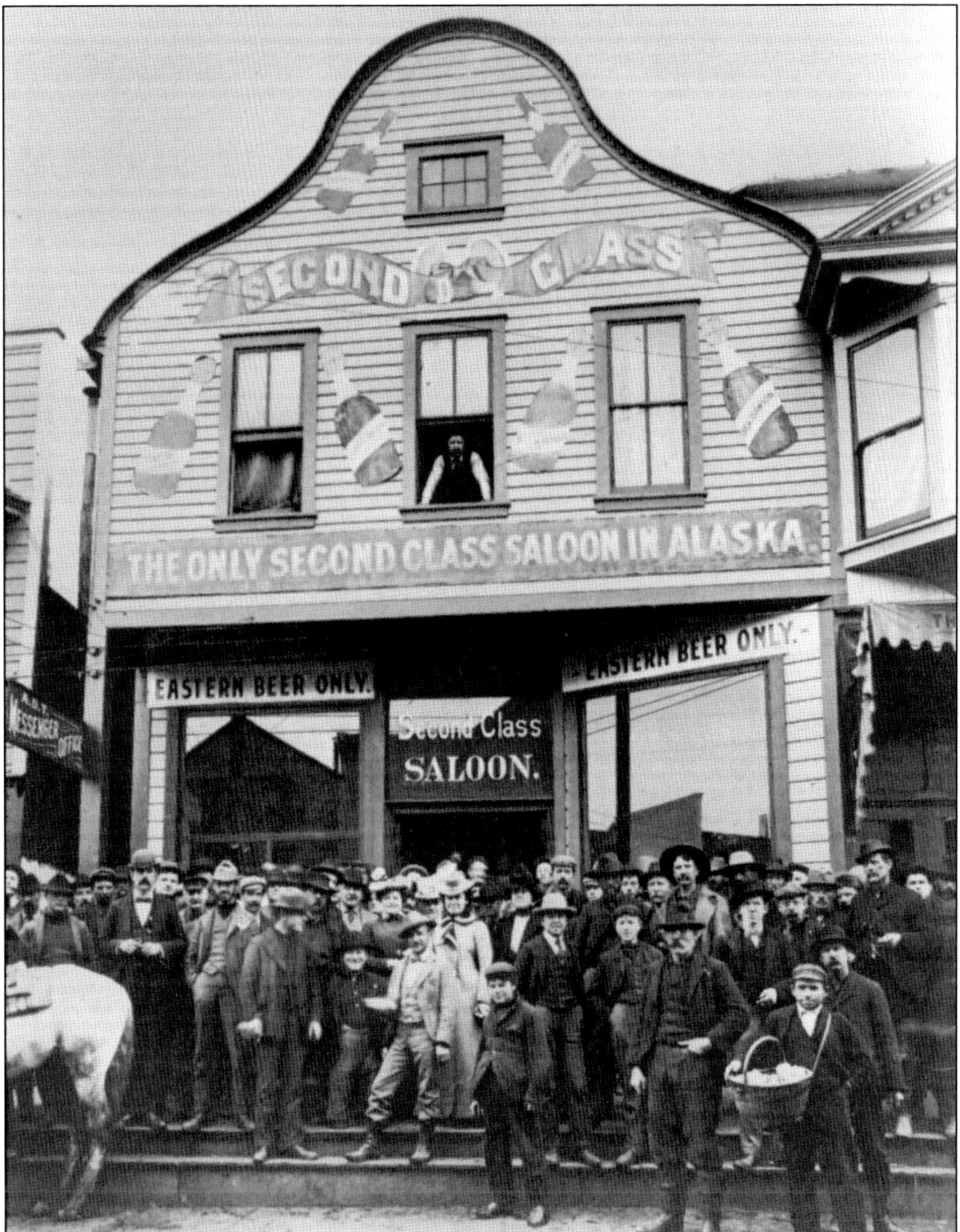

Dick Dawson and Charlie Suter ran the popular Second Class Saloon, which proudly served only Eastern beers. The saloon featured a 26-room hotel upstairs with carpets, electric call bells, and a convenient sewer line that extended from the rear of the building to the open sea. Saloon patrons in gold rush Nome often became unruly, as evidenced by a number of perpetrators who attempted to set the Second Class Saloon on fire for three consecutive days in 1900. (Carrie M. McLain Memorial Museum.)

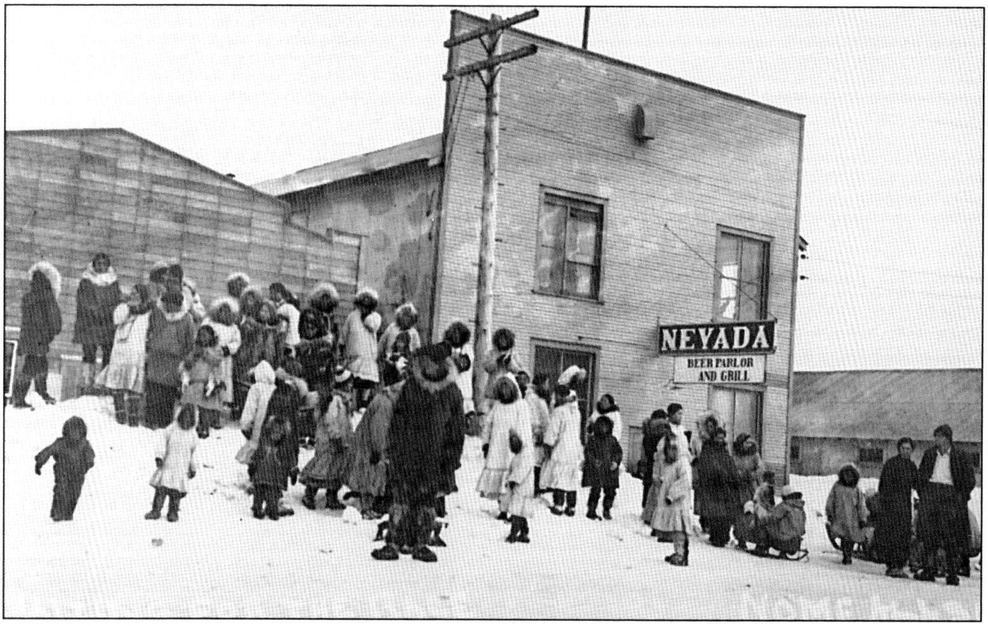

A crowd gathers outside the Nevada Beer Parlor and Grill in anticipation of the start of the All Alaska Sweepstakes dogsled race in Nome around 1915. The Nevada Bar served customers in various locations across town for over 40 years and became a popular watering hole for military personnel from Marks Air Force Base during World War II. (Carrie M. McLain Memorial Museum, NMP-83-1-218.)

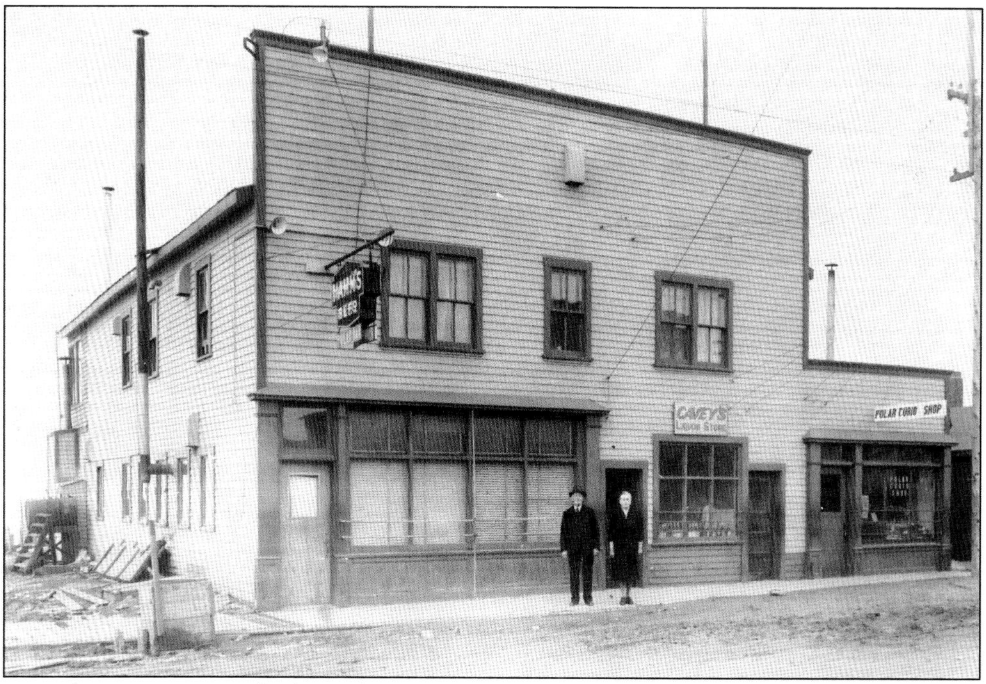

Nome business owners Charles and Wilhelmina (Witte) Cavey stand in front of their Polar Bar restaurant and liquor store during the 1940s. The restaurant on Front Street still dishes up hot food and panoramic views of Norton Sound. (Carrie M. McLain Memorial Museum, 2018.9.)

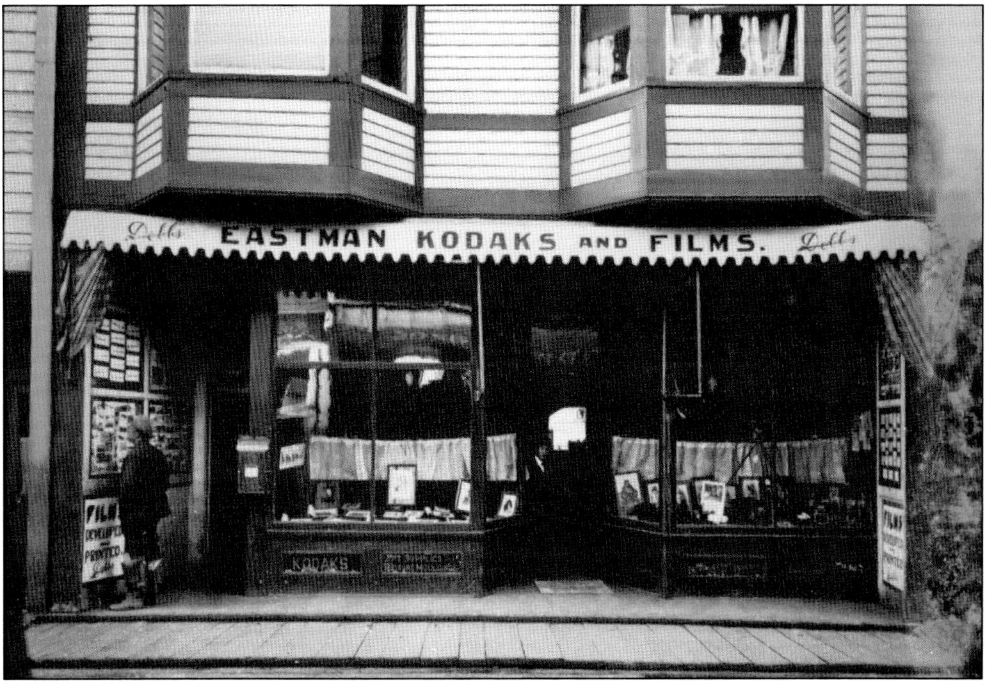

The Beverly B. Dobbs store on Front Street, seen here around 1908, sold art supplies and picture frames, shot studio portraits, and developed film from Kodak cameras. Dobbs took hundreds of photographs of Nome and founded a moving picture company that produced some of the first motion picture films north of the Arctic Circle. (Carrie M. McLain Memorial Museum, 97-9-40.)

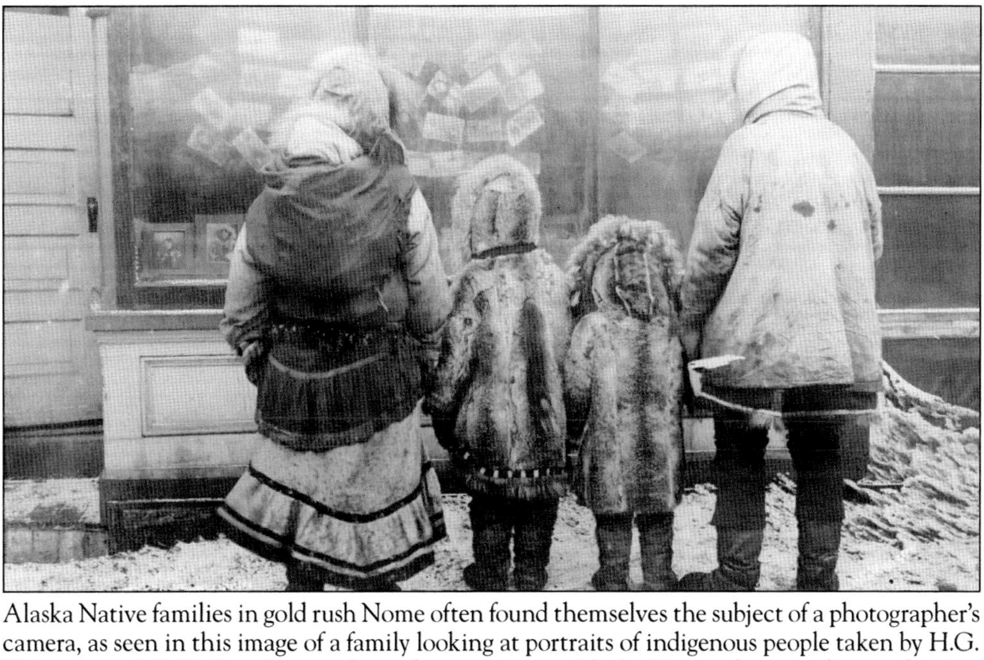

Alaska Native families in gold rush Nome often found themselves the subject of a photographer's camera, as seen in this image of a family looking at portraits of indigenous people taken by H.G. Kaiser around 1905. Growing numbers of contemporary Alaska Native photographers and artists are now producing visual representations of their culture and history on their own terms. (Carrie M. McLain Memorial Museum, McLain-628.)

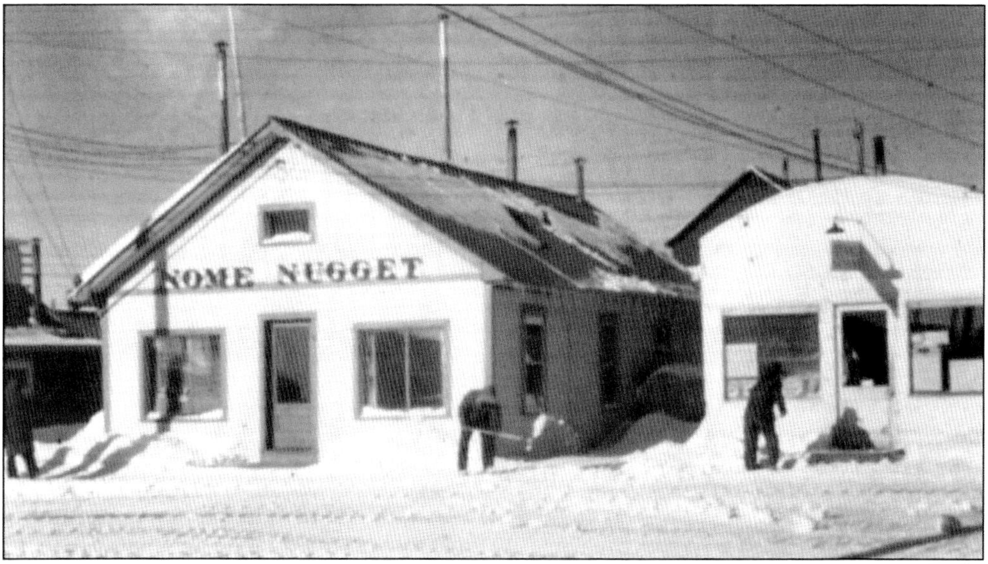

Workers shovel snow outside the *Nome Nugget* on Front Street after a winter storm during the 1950s. J.F.A. Strong founded the *Nome Nugget* in 1900 when newspapers functioned as profitable powerhouses of advocacy and sensationalism along with serious and objective news-gathering. The *Nome Nugget* continues to thrive as the only contemporary newspaper written and produced in northwest Alaska. (Carrie M. McLain Memorial Museum, 1998.4.17.)

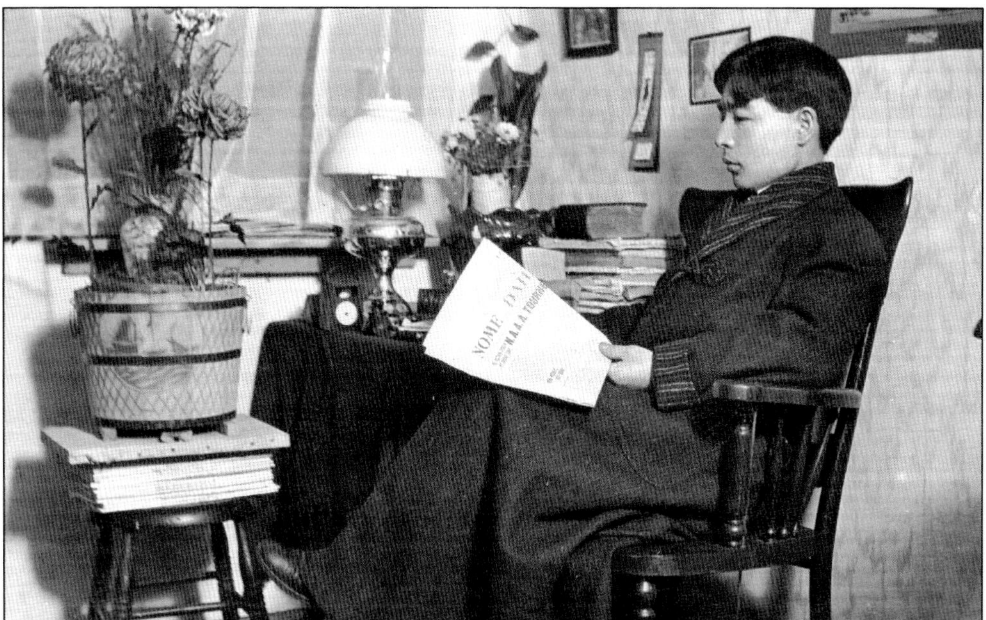

A young man relaxes in an armchair with the latest edition of the *Nome Daily News*, which ran in Nome from 1900 to 1904. Over 40 newspapers, including the *Nome Gold Digger*, *Cape Nome Rocker*, *Arctic Weekly Sun*, and *The Bering Straights*, have operated in Nome over the past 120 years. (Carrie M. McLain Memorial Museum, 1992.2.278.)

Seven

Resourceful Home Builders

The earliest structures of the Nome area were built by Inupiat families utilizing locally sourced driftwood, moss, and sod to create semisubterranean houses. Prospectors who arrived on the Nome beach in 1898 relied on walled canvas tents, but as the population grew, more substantial buildings were required. Steamships arrived in 1900 carrying architects, materials, and building plans for house styles to which newcomers were accustomed. Real estate developers met the sudden demand for single-family homes by constructing uniform houses in straight rows set a few blocks inland from the bustle of Front Street.

Houses in town featured tidy gardens behind picket fences and ornately furnished rooms with curtains in the windows. Families accepted high freight costs to ship prized pianos, sewing machines, and furniture to the city on the beach. Home interior spaces were decorated with Victorian style floral rugs and porcelain dishes alongside recent acquisitions of Alaska Native artwork and artifacts.

Keeping house in the early days of Nome posed a never-ending challenge. In the summer, strong winds blew dust and sand across the streets and through cracks around doors and windows. Before the luxury of indoor plumbing, bucket toilets known as "honey buckets" had to be carefully stored in preparation to be picked up and the contents carted away. In the winter, water had to be purchased from a horse-drawn delivery wagon and its use carefully monitored. Heavy snowfall tended to completely cover houses, and shoveling a path out the front door was often a daily routine.

Out on the tundra, miners faced more difficulties in constructing and maintaining suitable shelters. Cabins were typically constructed with driftwood or spruce logs brought in from the Council area and insulated with sod. Canvas tents provided semi–water repellent roofing, and wrecked ships offered a wealth of creative building uses.

In the summer, Nome homeowners enjoyed a wide selection of fresh produce and dairy products from local stores well-stocked with supplies from steamships, and from the gardens at Pilgrim Hot Springs, about 60 miles north of Nome. Access to outside produce ended with the autumn departure of the last ships from the roadstead. Families dried, smoked, canned, and stored food in barrels to try to make it last from October to June. When the fruit failed, home cooks turned apples into applesauce, oranges into marmalade, and bananas into bread.

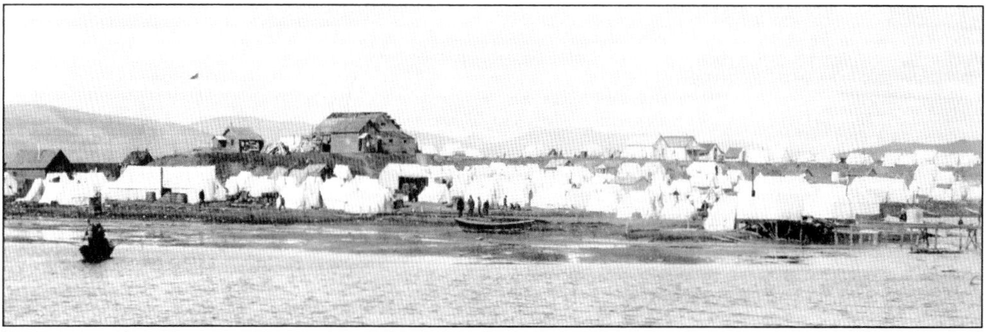

Early arrivals to Nome often set up their first home in a tent, as seen in this 1900 image of a crowded shoreline at Belmont Point, a couple miles west of the center of Nome. Bare rolling hills appear to the north while the Bering Sea stretches to the south and west. The footbridge on the right succumbed to the elements and was replaced by a suspension bridge in 1902. (Carrie M. McLain Memorial Museum, NMP-17-61.)

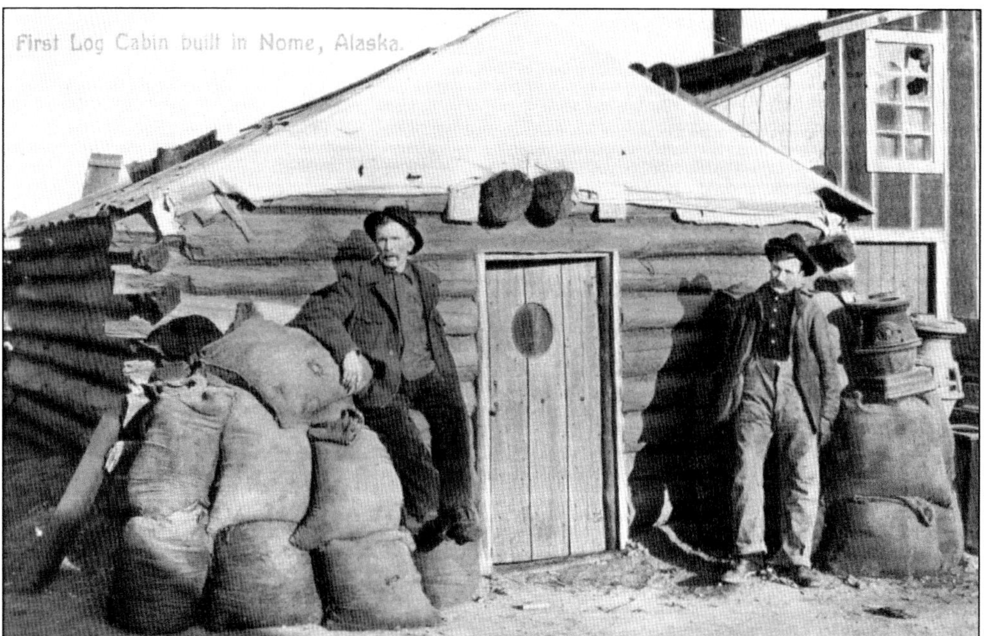

A scarcity of wood prompted ingenuity in the construction of houses, such as this tent-covered roof on the first log cabin built in Nome. Substantial lodging required homeowners to invest in lumber that arrived via steamship, or purchase logs that had to be transported over 70 miles from the sparse boreal forest around Council. (Carrie M. McLain Memorial Museum, 1979.1.437.)

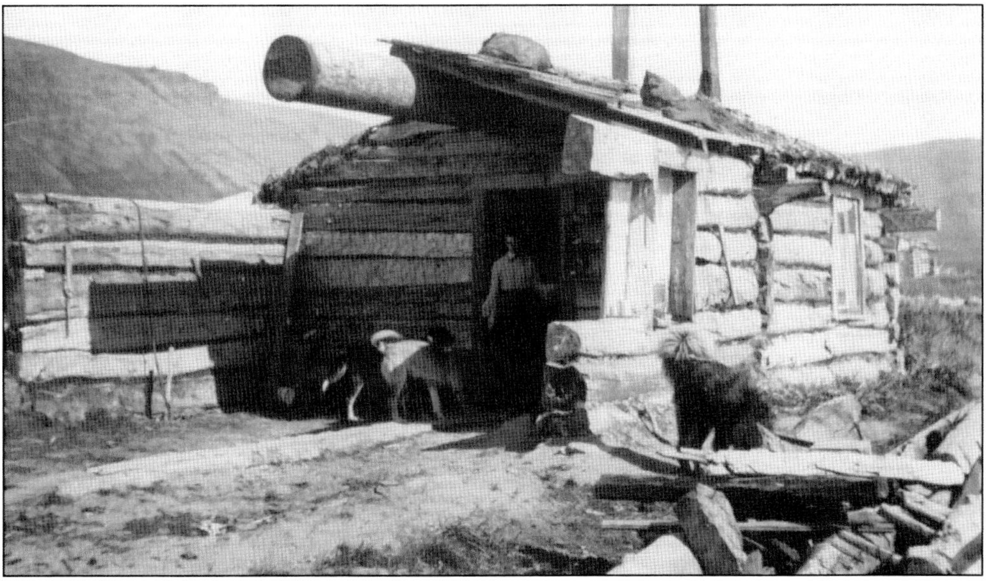

Jessie McDaniel pauses in the doorway of her Quartz Creek cabin, which was constructed from hewn logs, sod, and the mast of a wrecked ship, east of Nome. Jessie and her husband, Edmund, arrived in Nome with Edmund's brother Wilfred McDaniel in 1900 and spent the next six years mining gold, taking photographs, collecting Alaska Native artifacts, and documenting their experiences in colorful letters back home. (Photograph by Wilfred McDaniel; Carrie M. McLain Memorial Museum, MCD-521.)

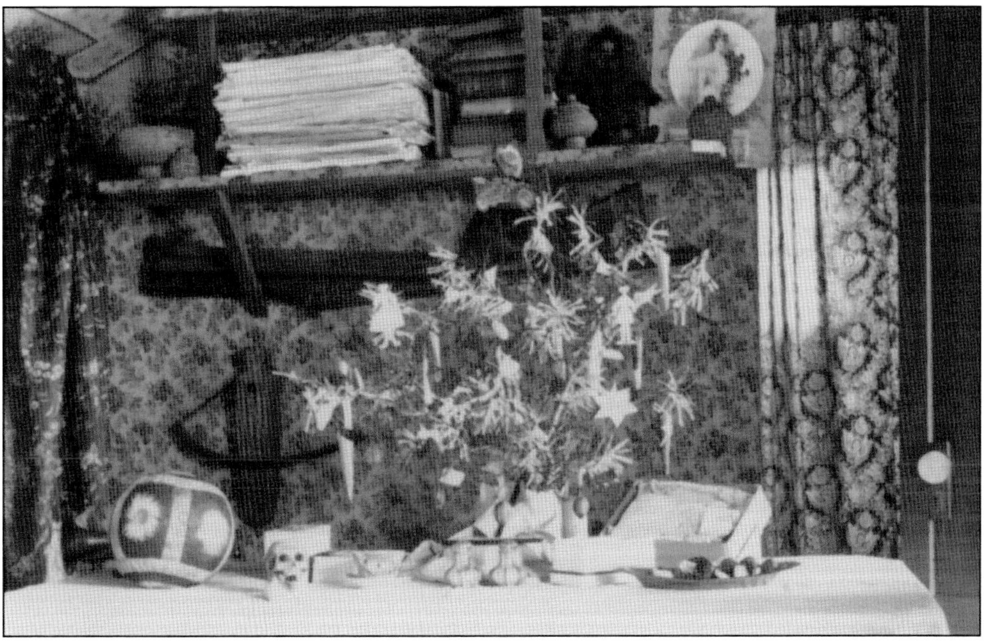

The McDaniel brothers outfitted a second cabin at Edwards Creek, shown here decorated for Christmas in 1906 with a small willow tree surrounded by walnuts, a reindeer skin ball, and a mechanical skull. The cabin offered a cheerful space during the dark winter with floral wallpaper, curtains, books, and comfortable furniture. (Photograph by Wilfred McDaniel; Carrie M. McLain Memorial Museum, MCD-229.)

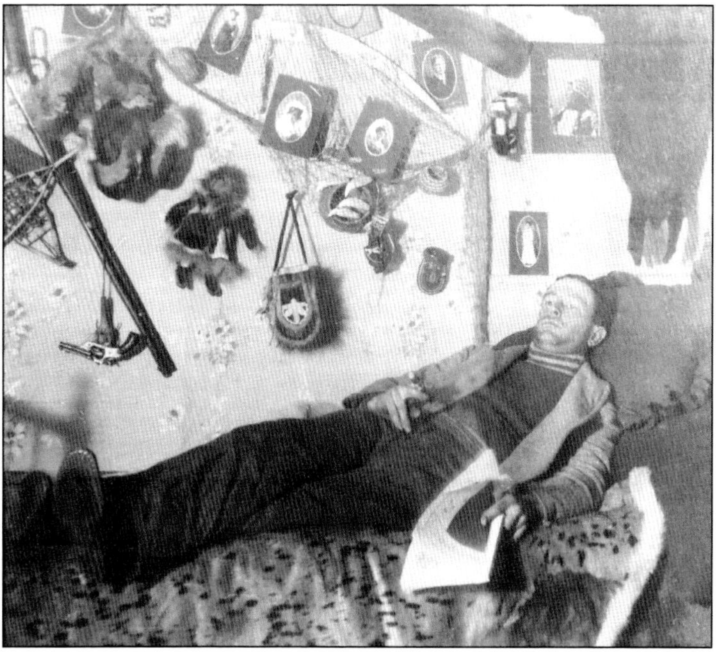

A Nome prospector relaxes on a spotted sealskin with a cigar and a copy of the mass-market magazine *Munsey* around 1905. Displayed against the floral wallpaper are an assortment of photographs, grass baskets, firearms, model snowshoes, a doll, an ivory cribbage board, and decorative storage pockets. (Carrie M. McLain Memorial Museum, Boyd-70.)

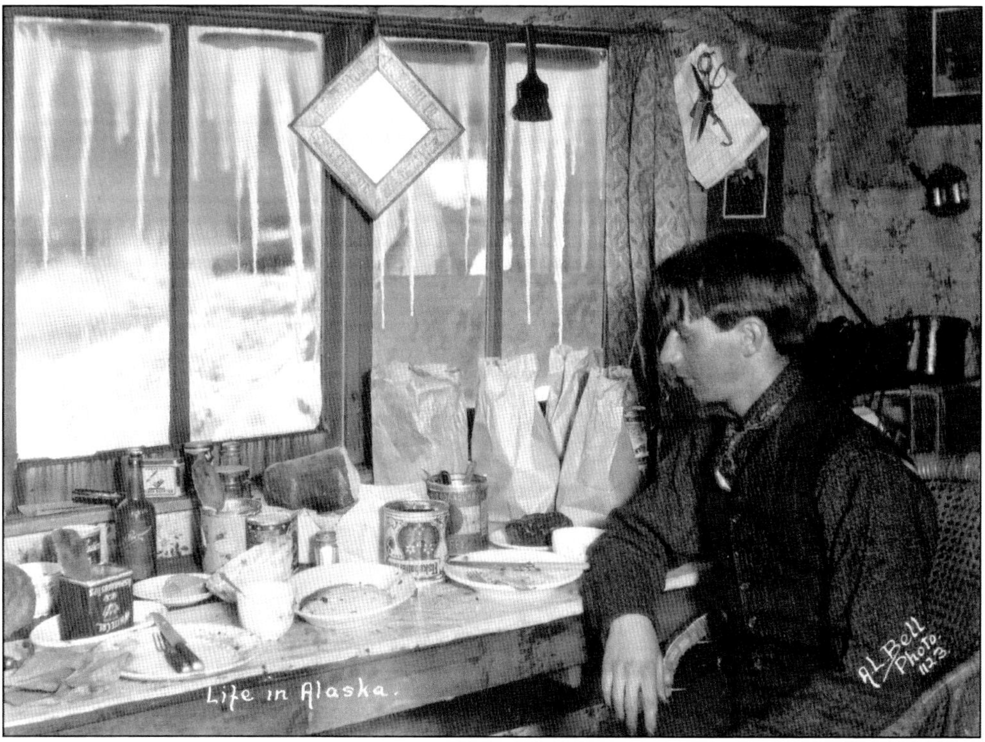

Photographer Arthur L. Bell stares dismally at a table of dirty dishes and winter provisions, comprising asparagus tips, Norwegian fiskeboller (fish balls), and Chase & Sanborn coffee. Bell trained as a photographer in his home country of Australia before traveling to Nome in 1903. He spent six years taking hundreds of images of the people, buildings, and activities in and around the gold rush town. (Carrie M. McLain Memorial Museum, Bell-N-116.)

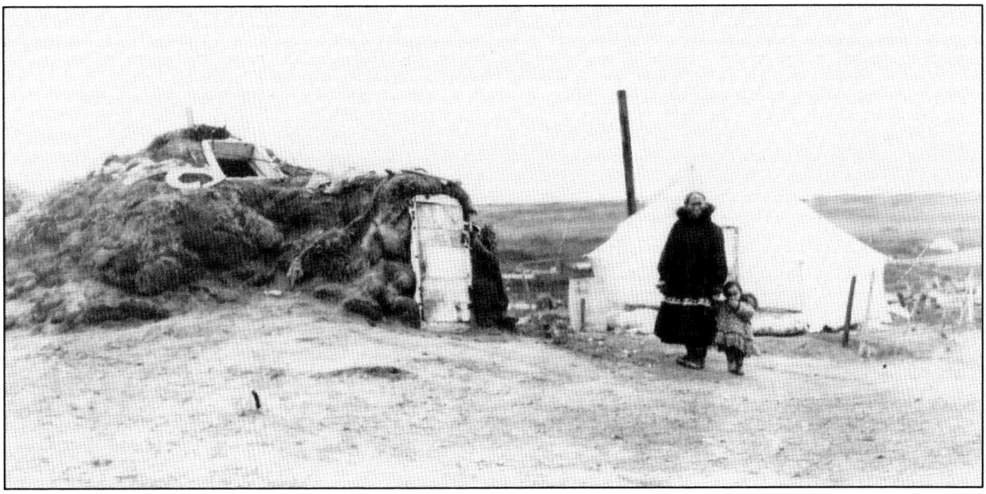

An Inupiat woman and child stand outside their summer tent and winter sod house on the tundra. A semisubterranean sod house (*igluuraq*) was built by thawing and excavating a large hole in the permafrost and erecting a framework made from timber or whale ribs over the hole. Wood walls and roofing were placed over the framework and then covered by a layer of moss and blocks of sod for insulation. Two partial sod houses were discovered during excavation of the Snake River Sandspit site in 2005–2006 that indicate successful use of this indigenous technology in the Nome area for at least 200 years. (Carrie M. McLain Memorial Museum, Woenne-789.)

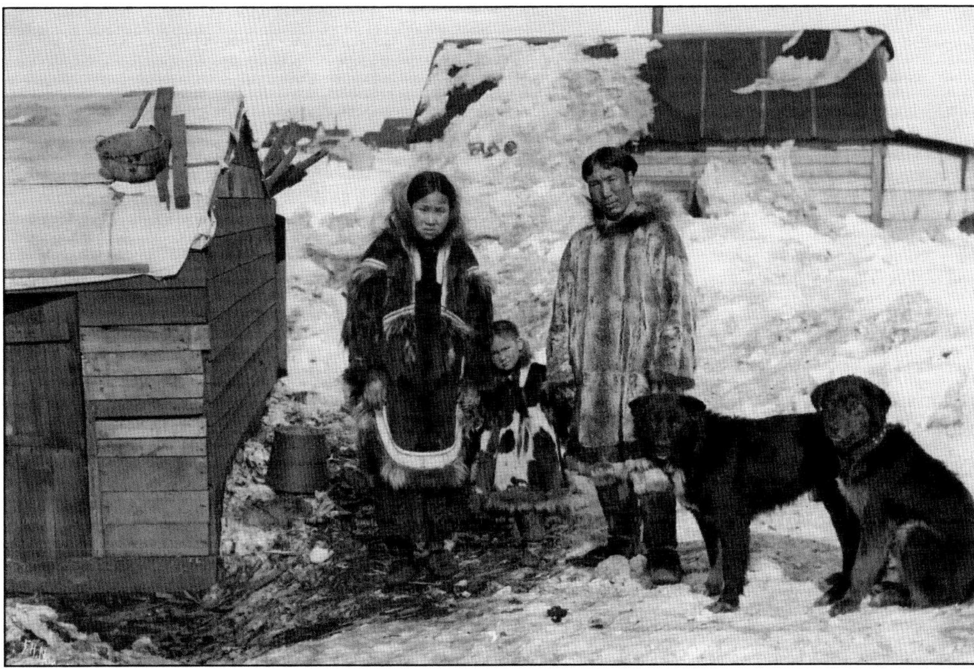

Celebrated ivory carver Angokwazhuk "Happy Jack" and his wife, Malinda, hold the hands of their daughter as they stand next to two sled dogs outside a snow-covered house in Nome around 1905. Happy Jack returned to Nome in 1900 and earned a substantial income as a professional artist, which allowed him to frequently purchase supplies and groceries from local stores and engage in subsistence activities for recreation. (Photograph by F.H. Nowell; Carrie M. McLain Memorial Museum, McLain-773.)

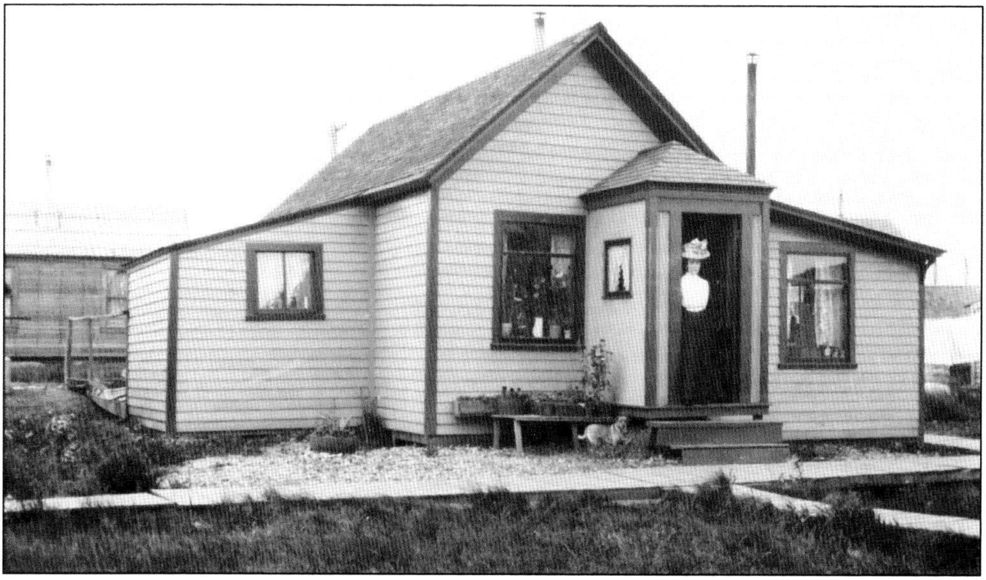

Nome residents brought existing ideas for building styles with them, as seen in this photograph taken between 1903 and 1909 of a simplified Queen Anne–style house with a steep roof and decorative front porch. The home has been spruced up for summer with curtains and flowers in the windows, a groomed lawn, and a swept sidewalk that holds back the unruly tundra. (Photograph by Arthur L. Bell; Carrie M. McLain Memorial Museum, Bell-N-475.)

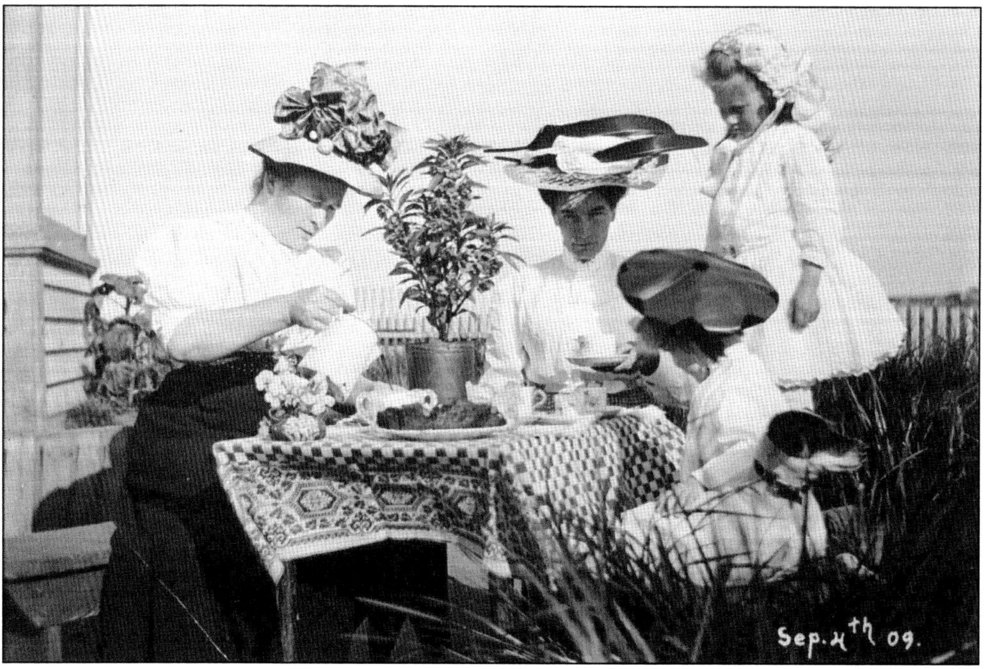

Clara Mielke stands on a garden bench to catch a better view of an afternoon tea table set with sweets and painted china on September 4, 1909. Nome residents continue to take advantage of the light-filled days of summer and cultivate a startling array of vegetables and flowers, which are shared along with growing tips during a popular garden tour each year. (Carrie M. McLain Memorial Museum, Mielke-1.)

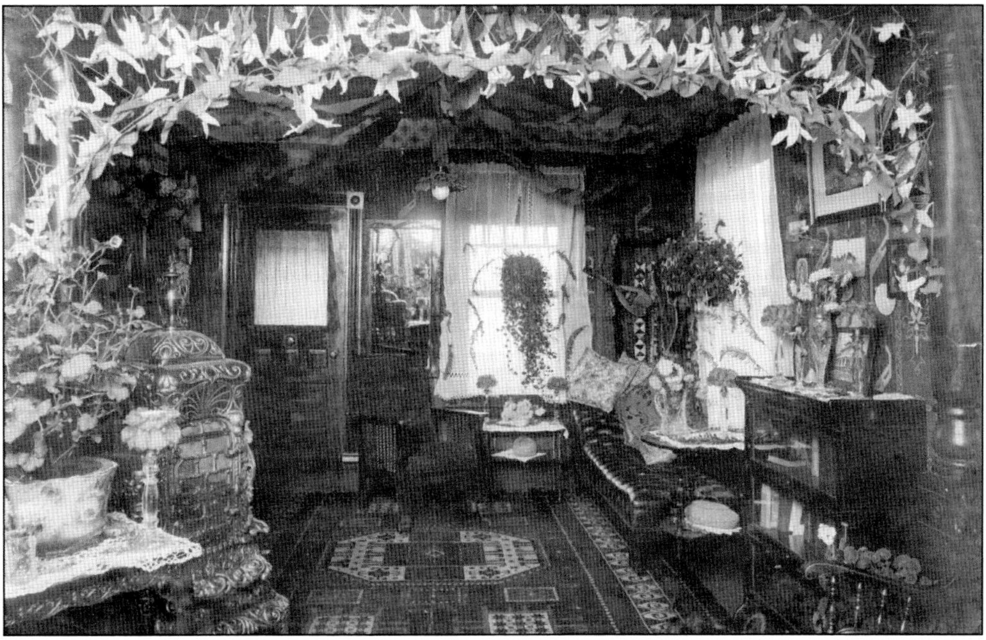

Paper lilies frame the home of W.L. Goodwin, superintendent of the Alaska Road Commission, in preparation for the marriage of his brother Frank Goodwin to Louise Glockler on October 2, 1909. The bride arrived in Nome aboard the SS *Victoria* the morning of the wedding. The ornate room combines Victorian floral motifs with local Alaska Native artwork, including coiled grass baskets and a reindeer-skin wall hanging. (Photograph by Alfred G. Simmer; Carrie M. McLain Memorial Museum, McLain-333.)

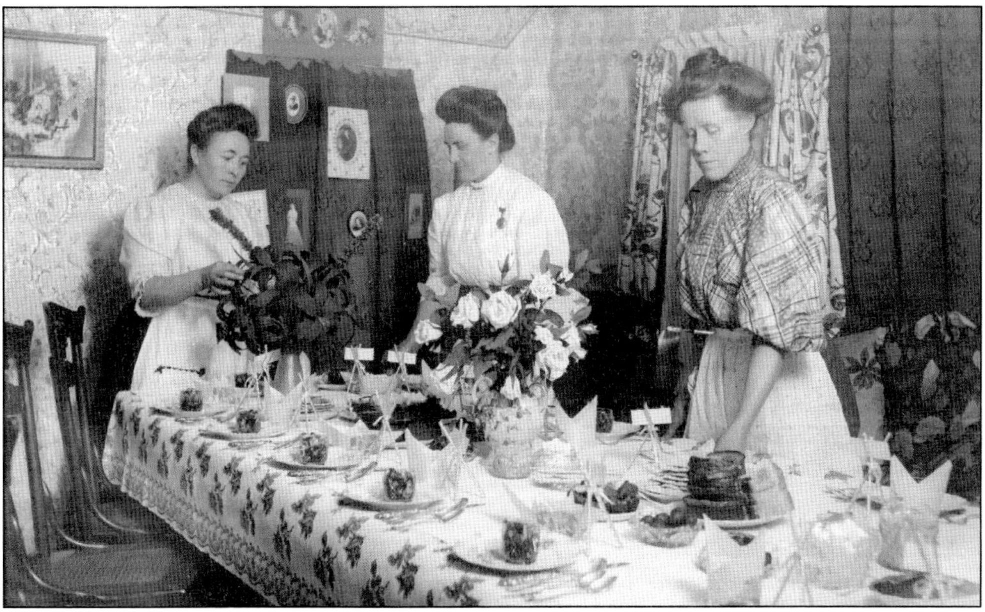

A group of women are setting the table for a fancy brunch with cut-glass dishes, prized fruit, and fresh flowers around 1905. Many women in Nome endeavored to create delicate beauty inside their homes despite the ruggedness of the landscape outside. (Carrie M. McLain Memorial Museum, 1979.1.207.)

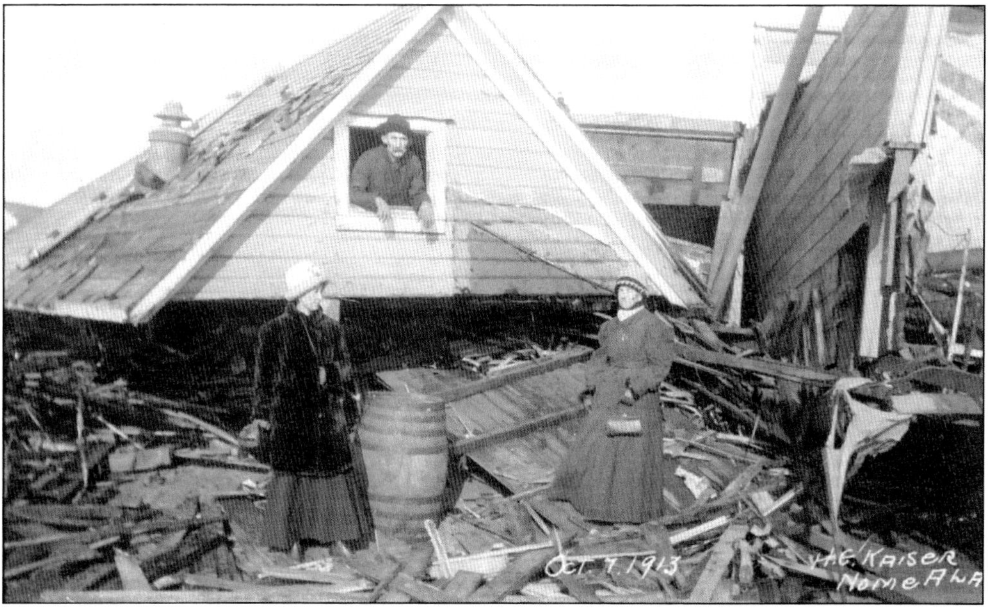

Nome residents often have to endure extreme weather conditions, as seen in this photograph of a home destroyed during the Bering Sea storm of October 1913. A series of fires and storms in the early 1900s prompted many residents to board steamships and head back to the Lower 48. (Photograph by H.G. Kaiser; Carrie M. McLain Memorial Museum, 1979.1.387.)

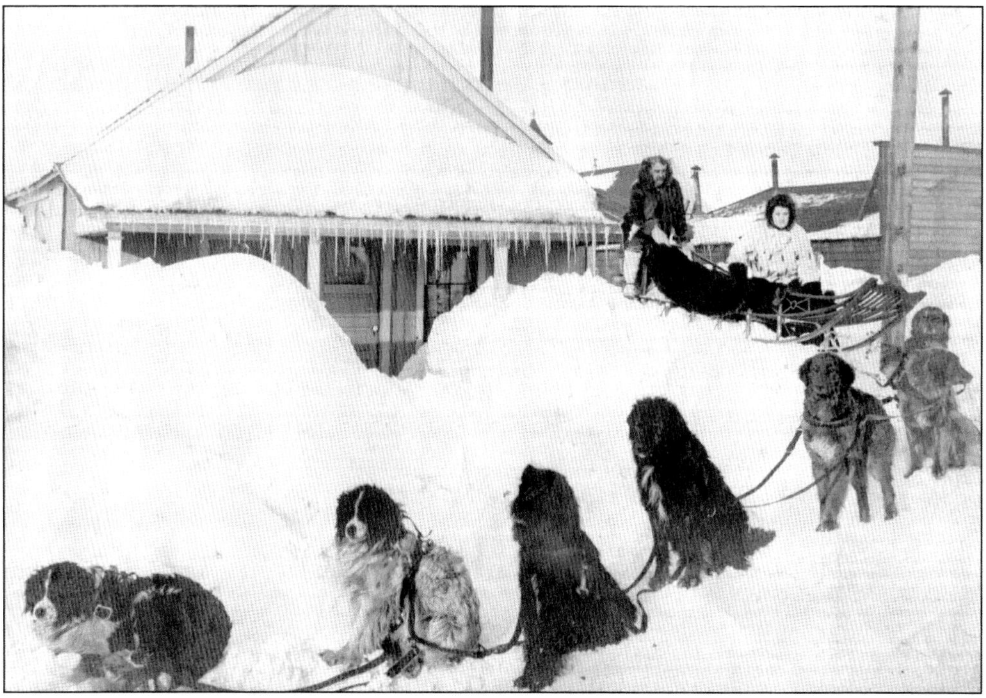

Heavy snowfall in the winter tended to cover houses in gold rush Nome until only a forest of stovepipes remained visible. Residents dug tunnels from their front doors, climbed out second-story windows, and used dog teams, like this couple, to escape the confines of their snow-clad homes. (Carrie M. McLain Memorial Museum.)

Eight

STAYING CONNECTED

Locked in by the ice-covered Bering Sea for seven months of the year, early Nome residents waited anxiously for steamships to bring fresh produce and outside news. The arrival of the first ship signaled a day of celebration as children were released from school, court was recessed, and families crowded the beach anxious for the appearance of relatives and supplies. Ships served as the main conduit to the outside world, and steamers from Seattle sometimes made up to 10 runs to Nome during the course of a season.

The water around Nome remains shallow for almost two miles out, which requires large ships to anchor in an area called the roadstead. Crossing the rough surf posed a perilous challenge that capsized small boats and encouraged creative solutions, such as the construction of a tramway scow built 600 feet out into the sea with a boxcar that lifted passengers over the waves. The majority of steamship passengers disembarked onto flat-bottomed barges called lighters and traveled over the surf until near the shore, where a gangplank completed the distance. Lighterage companies enjoyed hearty business, and during the summer of 1900, transported over 15,000 passengers and 600,000 tons of freight to the Nome beach.

The Nome roadstead and the mouth of the Snake River presented a continually changing scene, with steamers and ships at anchor, tugs and gasoline schooners pulling lighters, sternwheelers and small boats plying the Snake River, and skin boats arriving at the Sandspit. Once on land, horse-drawn wagons and narrow-gauge railroads shuttled people and supplies to mining claims across the Seward Peninsula. Far-flung mining camps enjoyed the convenience of an early long-distance telephone system that connected them to the center of Nome. By 1910, pedestrians and horses were competing with automobiles chugging down Nome's narrow streets, and by the late 1920s, airplanes had largely replaced dog teams as the major means of long-distance travel. Although steamships no longer bring thousands of gold-seekers to Nome, the Port of Nome has remained vital to connecting people and activities across the Arctic. Freight barges, commercial fishermen, scientific expeditions, and passenger cruise ships all visit Nome to explore the historic town, move supplies, stage equipment, and resupply their vessels.

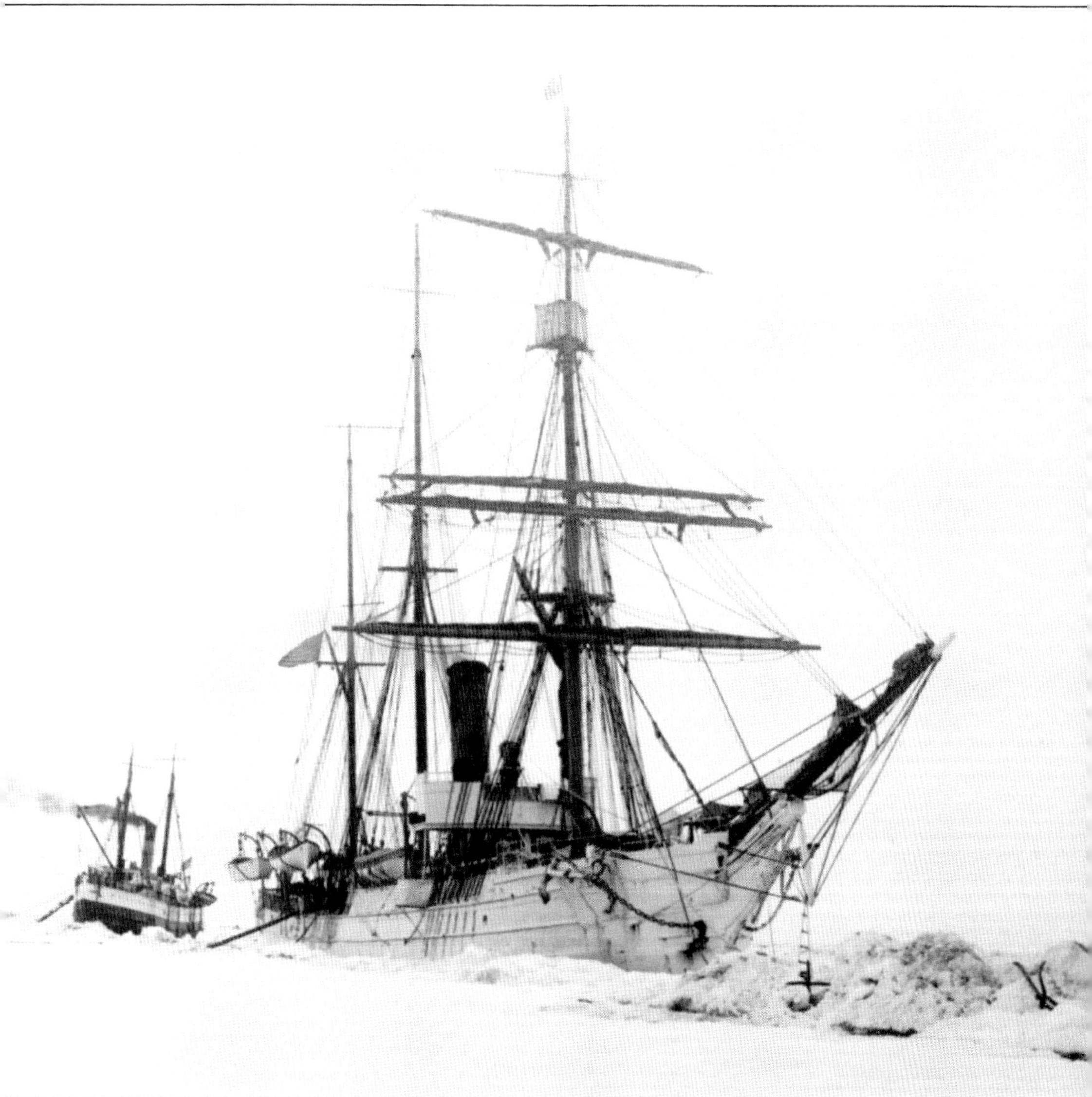

The US revenue cutter *Bear* (front) and the SS *Corwin* (behind) are shown anchored in the Nome roadstead around 1910. The *Bear* was built in 1874 as a dual steam-powered and sailing ship with six-inch sides that made her a forerunner of modern icebreakers. From 1885 to 1926, the *Bear* sailed from her port in Oakland, California, north to Alaska, where she cruised the Bering Sea on the lookout for seal poachers, shipwrecked whalers, and illegal trade with Alaska Natives. Long-term captain Michael Healey became renowned for ferrying reindeer from Siberia to Alaska, transporting lumber to build remote schools across the Bering Sea, and delivering supplies and medical support to Alaska Native communities. (Photograph by Lomen Brothers; Carrie M. McLain Memorial Museum, NMP-82-25-2.)

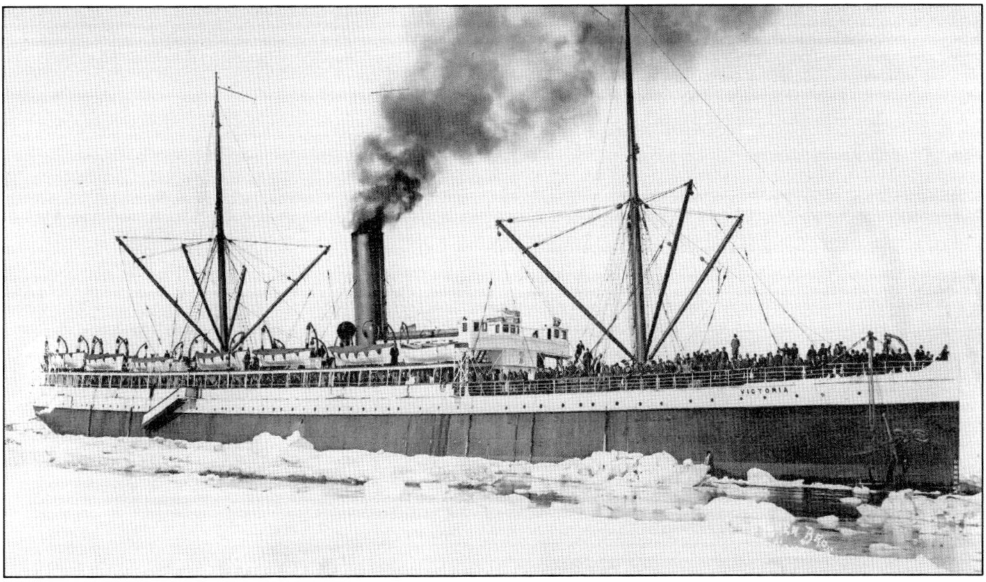

The SS *Victoria*, seen here in 1912, first sailed to Nome in June 1900 with over 1,000 passengers and 200 tons of freight. In October 1918, the *Victoria* left Seattle and carried the devastating Spanish flu to Nome. The influenza epidemic swept across the Seward Peninsula and killed over 1,000 people in the Nome area. The Catholic Church opened an orphanage at Pilgrim Hot Springs and cared for almost 125 Alaska Native children whose parents and families were lost to the epidemic. The *Victoria* received special government permission to increase her load-carrying capacity and transported 603 survivors out of Nome and back to Seattle. (Photograph by Lomen Brothers; Carrie M. McLain Memorial Museum, McLain-833.)

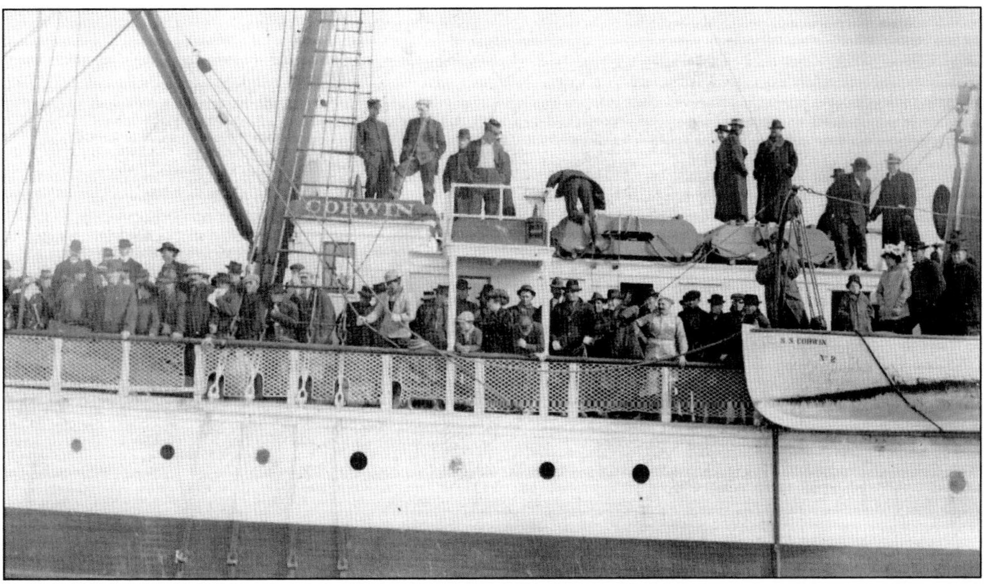

Passengers crowd the deck of the SS *Corwin* on her arrival into Nome in 1903. The *Corwin* was regularly one of the first steamers to reach Nome in the spring and one of the last to leave in the fall. In 1904, the *Corwin* underwent extensive modifications, including the addition of electric lighting throughout the ship and running water in all the staterooms. (Carrie M. McLain Memorial Museum, McLain-851.)

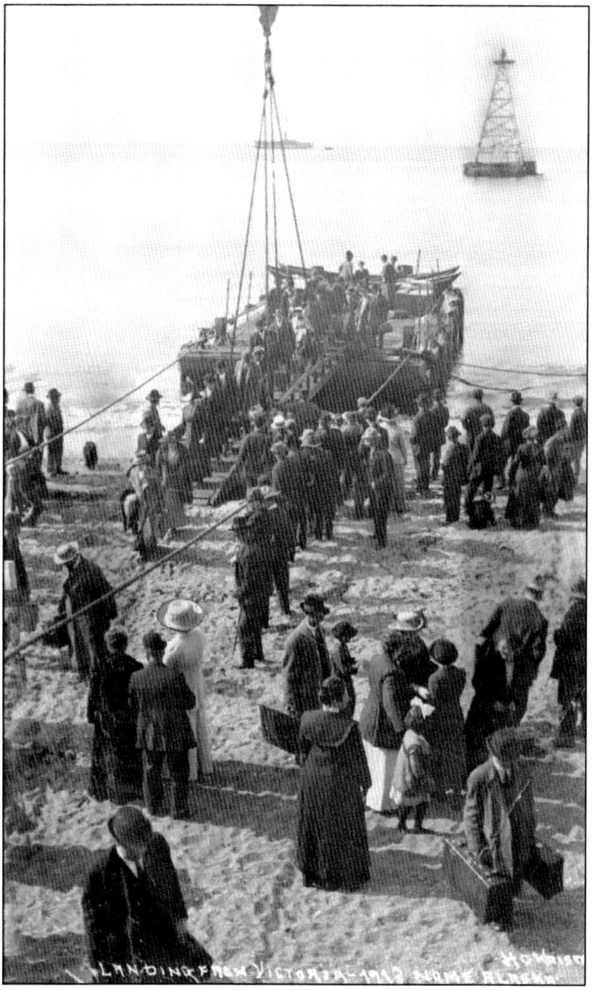

Family members and friends greet newly arrived passengers from the SS *Victoria* in 1913. The image shows a lighter that has drawn close to the shore and under a long boom, which has lowered a gangplank onto the beach. A tramway scow can be seen in the middle distance. (Photograph by H.G. Kaiser; Carrie M. McLain Memorial Museum, McLain-869.)

Steamship passengers carry suitcases, boxes, and bedding across a gangplank on the final leg of their journey to Nome. Lightering services, such as the John J. Sesnon Lighterage Company and Lomen Commercial Company, enjoyed steady business transporting people and freight over the rough surf of the Bering Sea. (Photograph by Otto D. Goetze; Carrie M. McLain Memorial Museum, Rolando-35.)

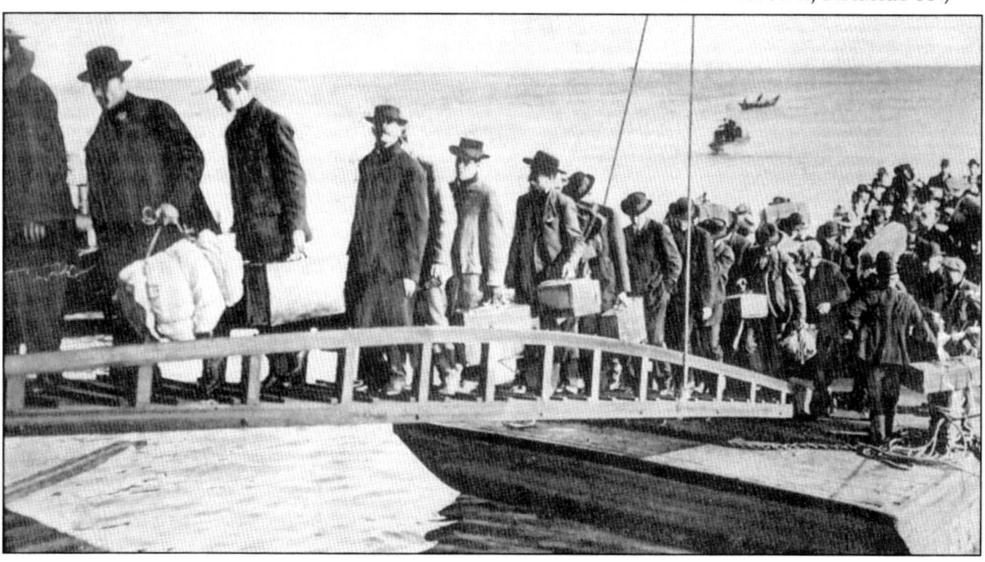

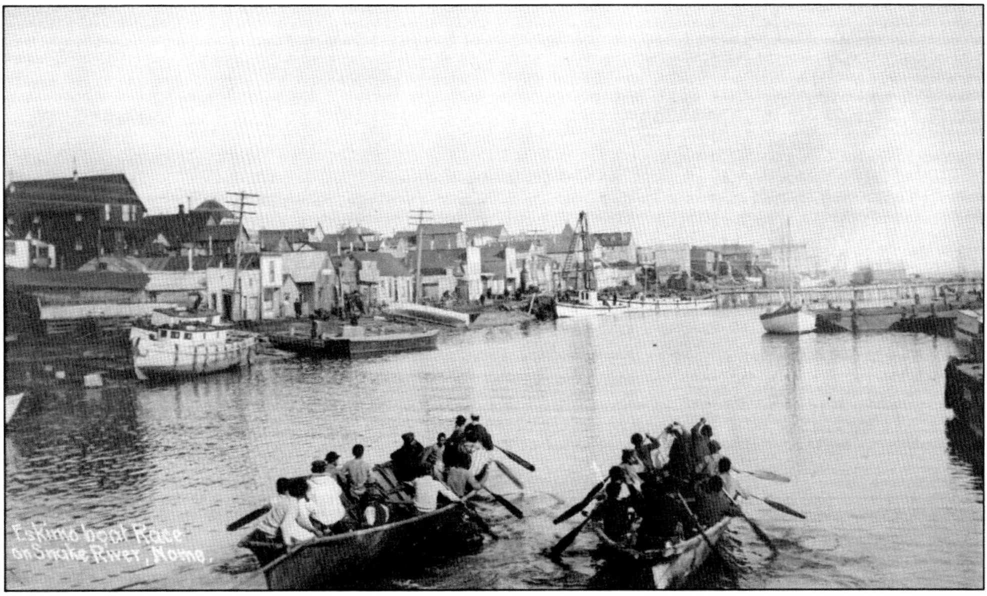

This photograph shows two skin boats participating in a race on the Snake River around 1905. A skin boat (*umiaq*) consists of a light driftwood frame over which a bearded seal or walrus hide cover is stretched. Traditional skin boats ranged from 20 to 50 feet in length and were measured by the number of skins required to cover the sides. A 30 foot traveling boat could carry over 10 passengers and 3,000 pounds of freight. (Carrie M. McLain Memorial Museum, NMP-4-165.)

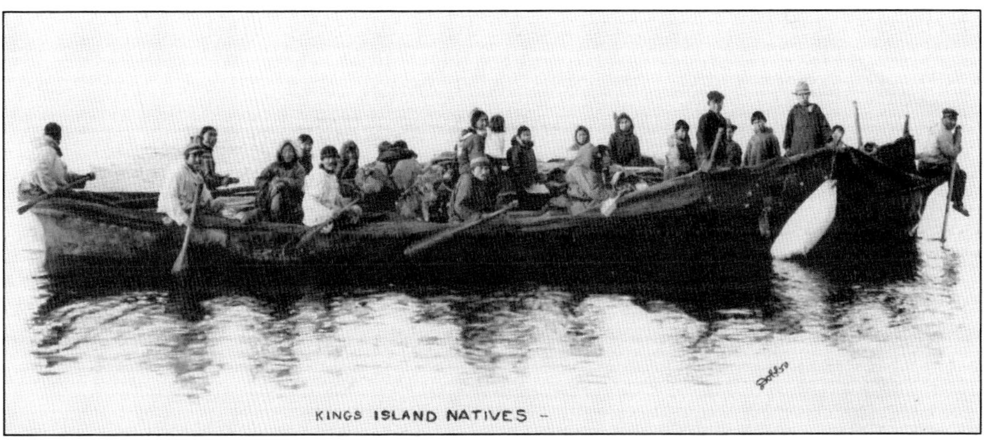

Travelers from King Island use driftwood paddles to propel this large skin boat that has an inflated sealskin float lashed to the bow for stability around 1900. A skin tarp (*quliutaq*) is wrapped around the boat and held up with sticks to keep water and wind off the passengers. (Photograph by B.B. Dobbs; Carrie M. McLain Memorial Museum, 97.2.113.)

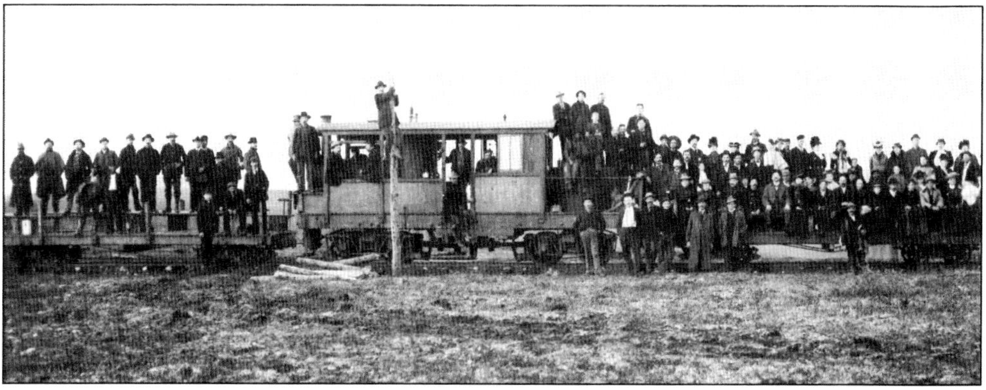

Passengers swarm the Wild Goose Railroad, built by Charles D. Lane in 1900 to haul people, freight, and supplies from the mouth of the Snake River to mining claims along Anvil Creek. The five miles of narrow gauge railroad terminated at Discovery Station; due to a lack of a turntable, it made the return trip to Nome backwards. (Carrie M. McLain Memorial Museum, NMP-80-3-520.)

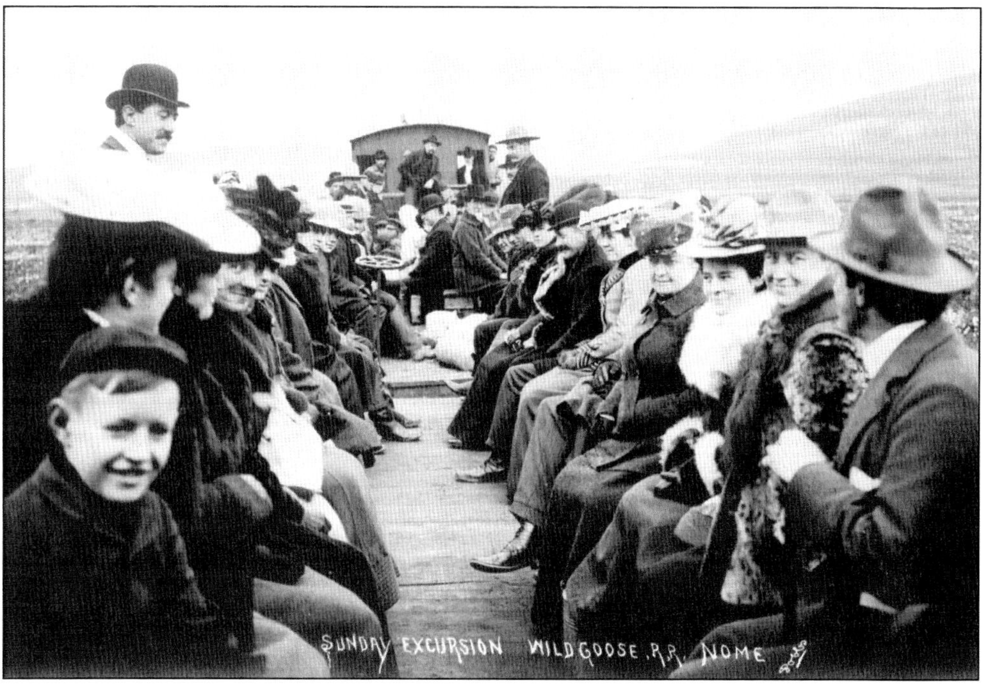

Well-dressed passengers smile at photographer B.B. Dobbs during a Sunday excursion aboard a boxcar on the Wild Goose Railroad around 1900. Customers could board the train at the Snake River Sandspit and travel the wavy track at a pace of eight miles per hour for 50¢ each way. (Photograph by B.B. Dobbs; Carrie M. McLain Memorial Museum, 96.26.27.)

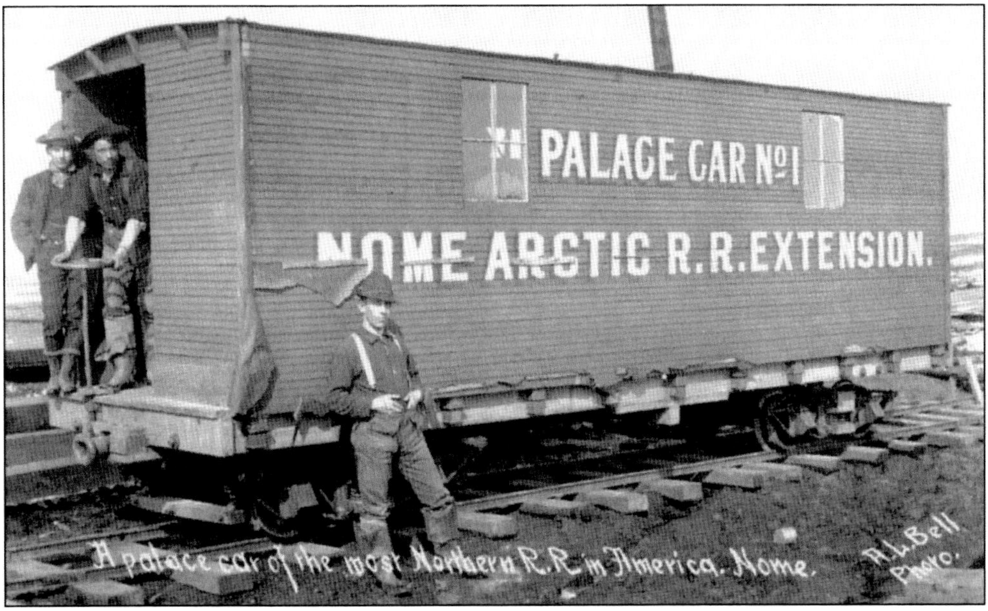

The Wild Goose Railroad was reorganized as the Nome-Arctic Railroad in 1903. Owners extended the tracks inland to Dexter Station in 1904, but high taxes and labor problems prompted closure of the little railroad after only four months of operation. (Photograph by Arthur L. Bell; Carrie M. McLain Memorial Museum, 1999.2.566.)

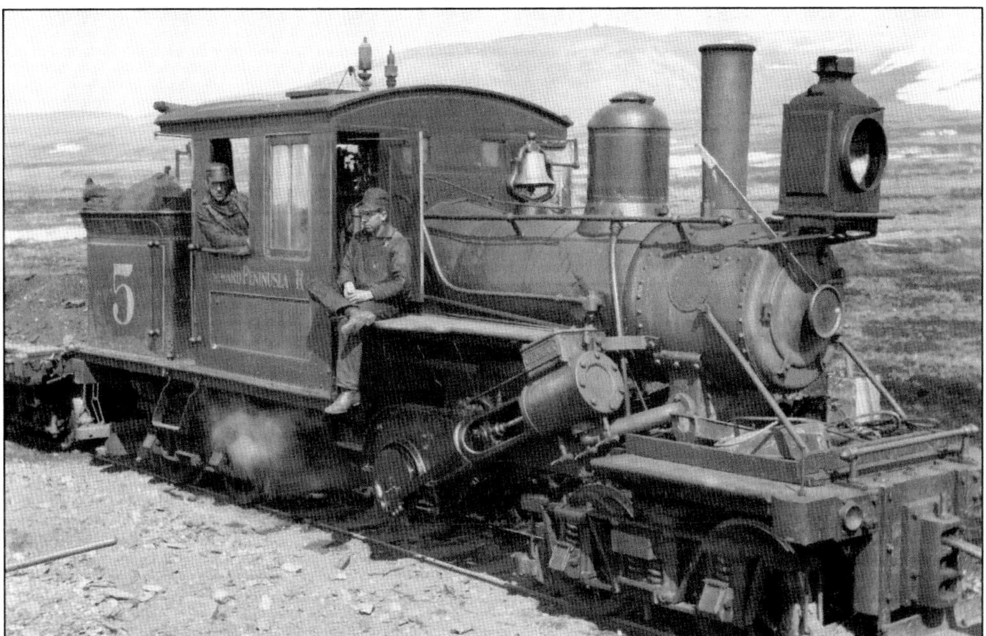

The Seward Peninsula Railroad incorporated in 1906, and during that summer, the rail line was extended from Dexter to the Kuzitrin River on the southern edge of the Kougarok mining district. The railroad terminated at Shelton, which grew to include a store, roadhouse, and a post office, along with two daily scheduled runs to Nome. (Photograph by Arthur L. Bell; Carrie M. McLain Memorial Museum, Bell-N-564.)

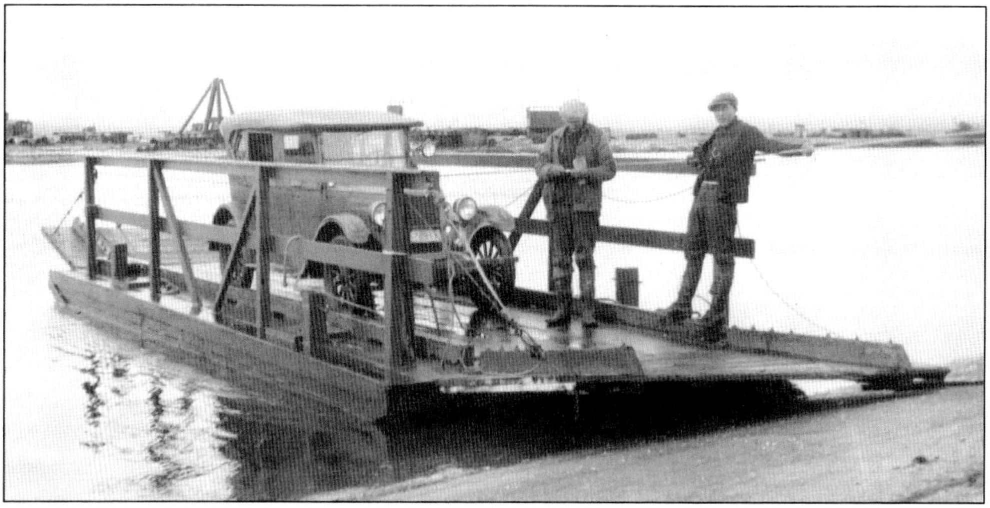

An automobile driver studies a slip of paper as his vehicle is ferried across Safety Lagoon around 1930. Charley Dalquist from Stockholm, Sweden, became proprietor of Safety Road House in 1910 and ran the ferry there until 1951. The ferry transported horses and dog teams in the days before cars dominated the Nome roadways. (Carrie M. McLain Memorial Museum, 1979.1.586.)

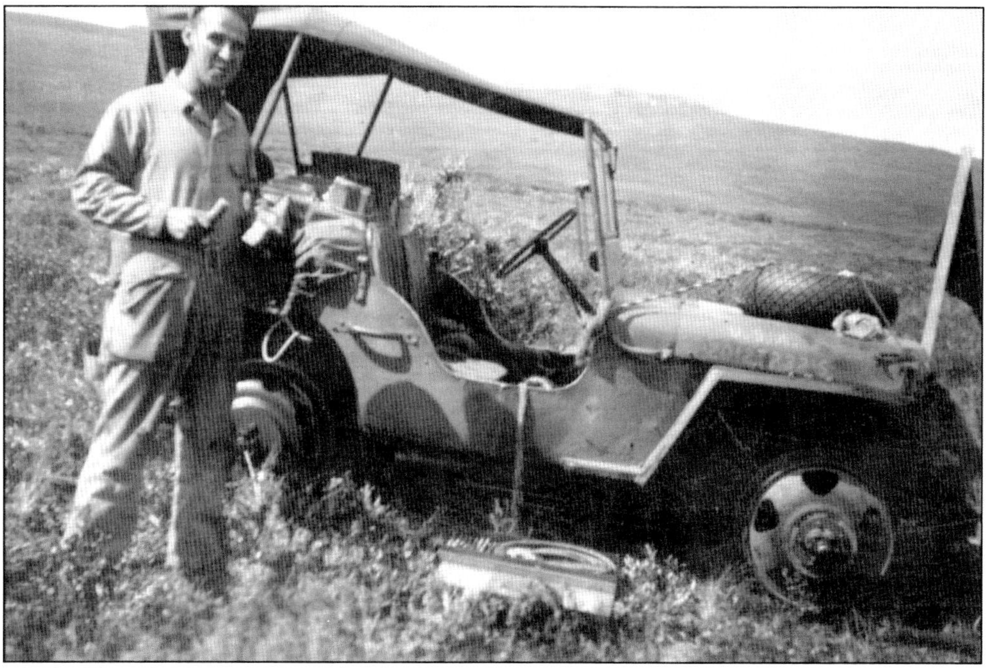

The buildup of military personnel and supplies at Marks Air Force Base during the 1940s brought an assortment of utility vehicles to Nome. Durable and off-road-capable jeeps offered a welcome opportunity to explore the rolling tundra. (Carrie M. McLain Memorial Museum, 2000.16.80.)

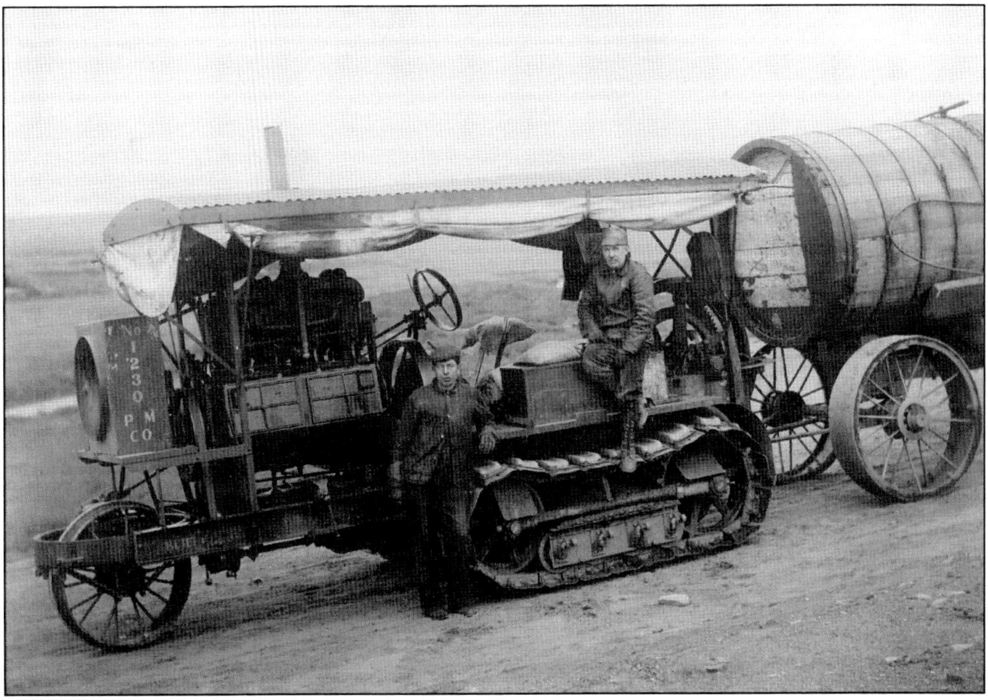

A Holdt Caterpillar, operated by the Pioneer Mining Company, pulls a barrel of drinking water to Nome around 1935. Early tractors were prone to pop wheelies and flip backwards, crushing their operators. The addition of caterpillar-style tracks added weight and helped reduce potential accidents. (Carrie M. McLain Memorial Museum, McLain-343.)

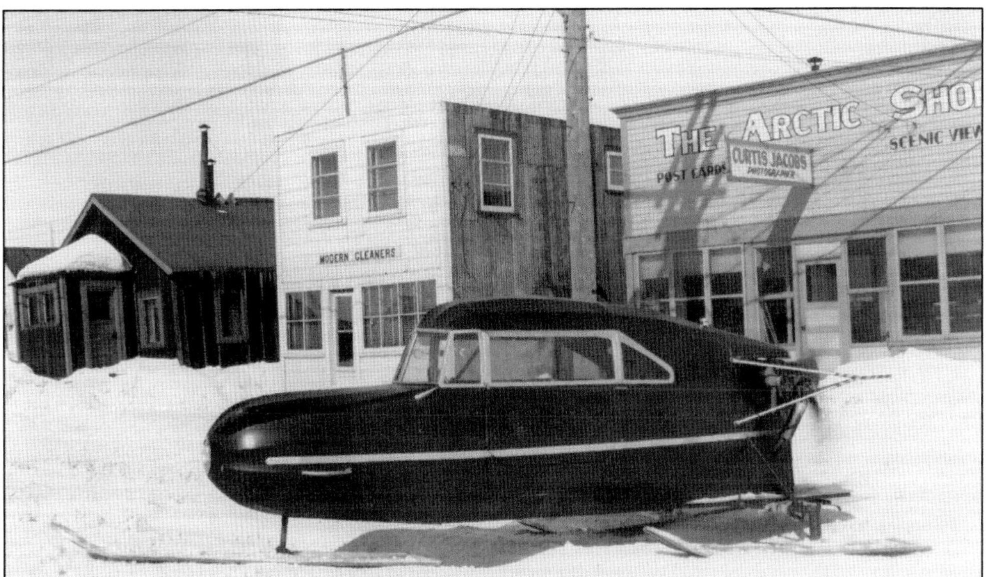

A hybrid snowmachine with skis and a propeller buzzes outside the Arctic Shop, run by photographer Curtis Jacobs, on Front Street, seen here around 1935. Over the years, Nome residents have navigated the deep snowdrifts with dogsleds, skis, snowshoes, snow machines, and an assortment of vehicles outfitted with skis or tracks. (Carrie M. McLain Memorial Museum, Jacobs-744.)

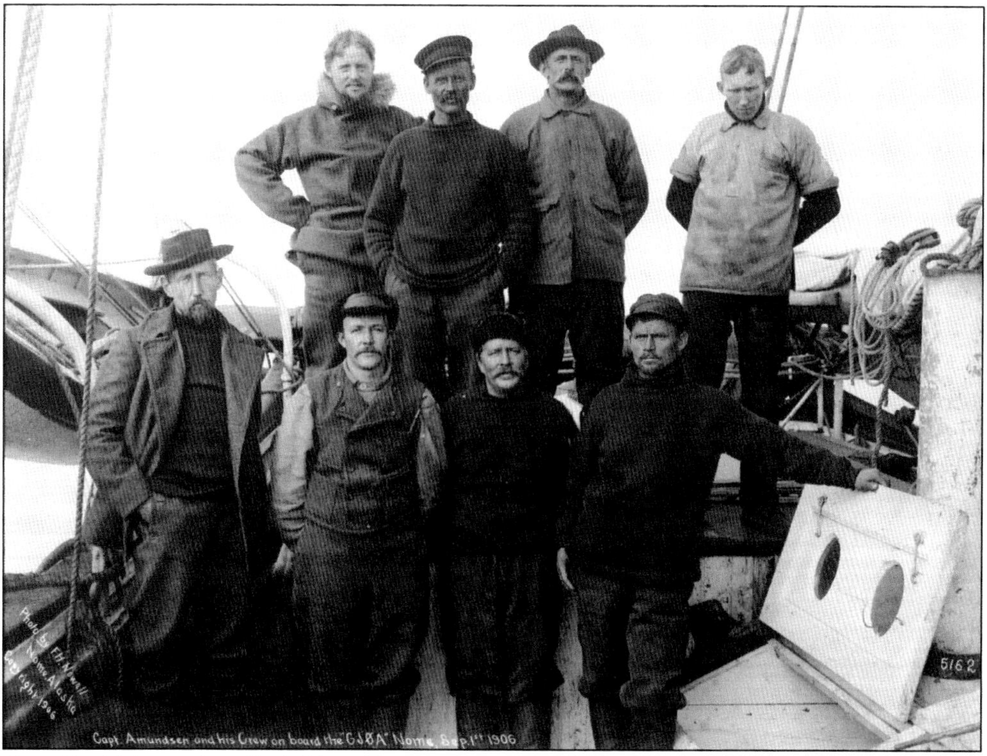

Capt. Roald Amundsen (far left) and his crew are pictured aboard the *Gjoa* in Nome on September 1, 1906. The *Gjoa*, a small 47-ton sloop, was the first ship to navigate the Northwest Passage, and its crew spent over three years gathering scientific data on the magnetic North Pole. Amundsen and his crew were hailed as heroes when they reached Nome and were honored with banquets and a parade. (Photograph by F.H. Nowell; Carrie M. McLain Memorial Museum, NMP-23-11.)

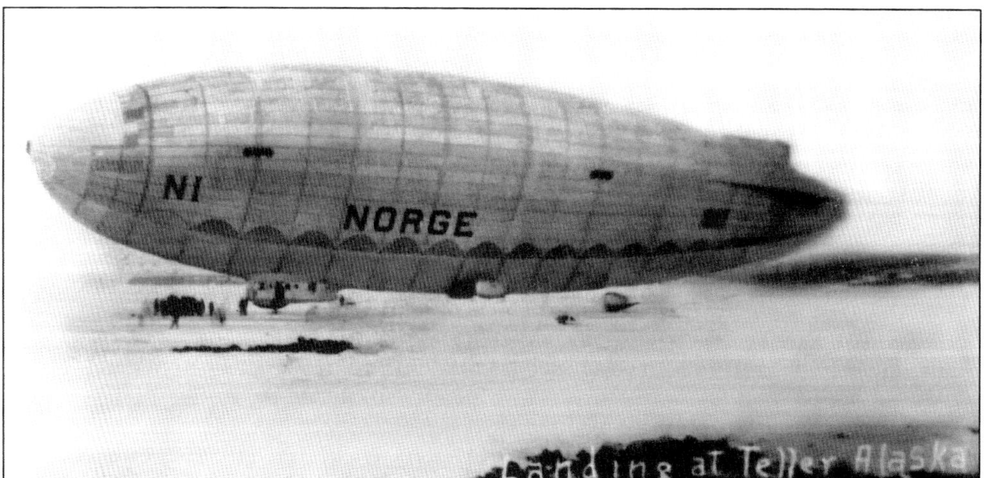

Expedition leader Roald Amundsen guided the airship *Norge* on the first successful flight to the North Pole in 1926. Although scheduled to complete the trip in Nome, the *Norge* landed on the ice 72 miles northwest at Teller due to poor weather conditions. (Carrie M. McLain Memorial Museum, NMP-95-3-6.)

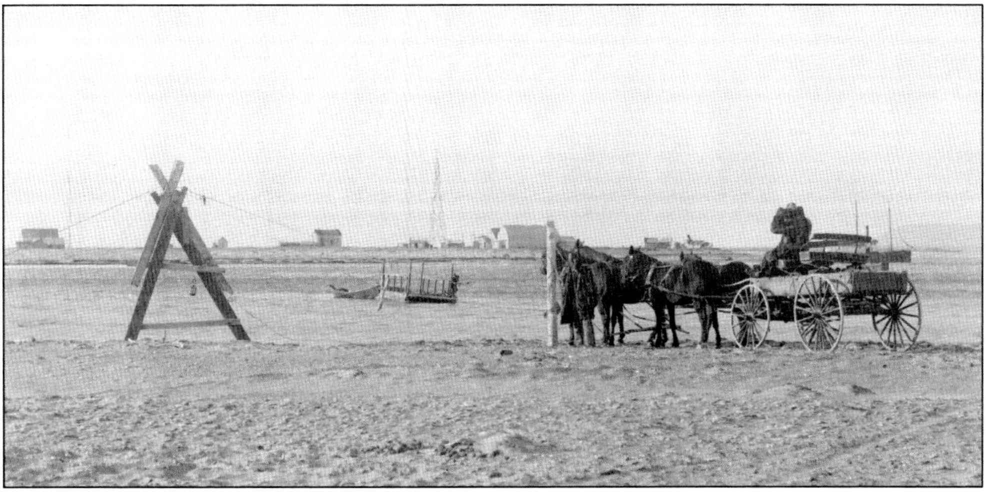

A horse team waits to board the ferry next to telegraph lines at Port Safety on October 2, 1905. The US Army Signal Corps led construction of the Washington Alaska Military Cable and Telegraph Service (WAMCATS) from 1900 to 1905. The system connected military headquarters across Alaska with a network of undersea cables and landlines that stretched 1,500 miles across glacial streams, boggy tundra, and mountainous terrain. The line from Nome extended 24 miles east via Fort Davis to Safety Harbor and was completed in the summer of 1900. (Photograph by F.H. Nowell; Carrie M. McLain Memorial Museum, NMP-20-768.)

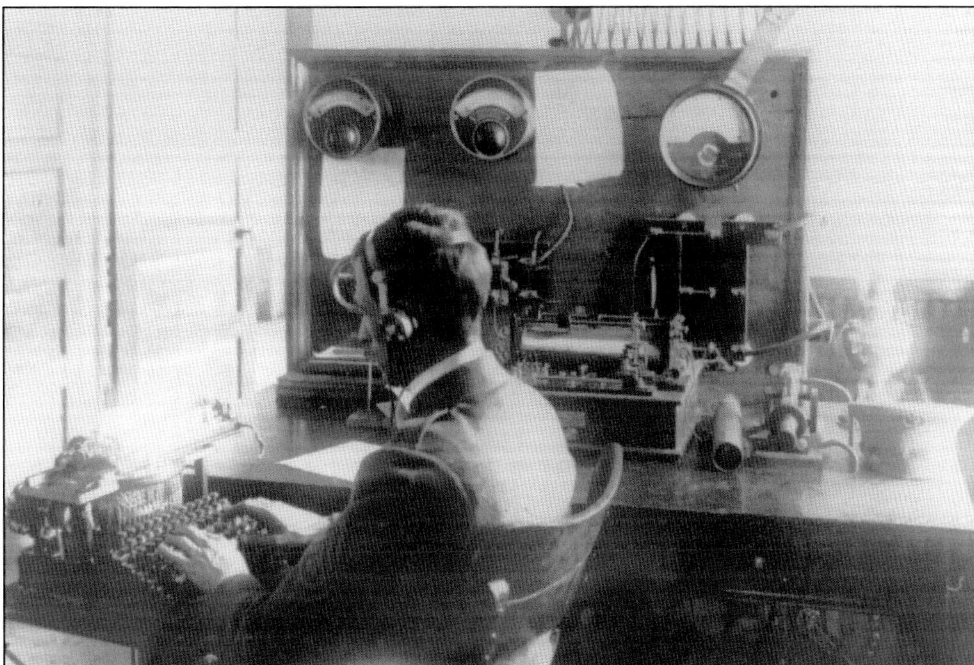

The WAMCATS system linked the residents of Nome to the outside world and made it possible to quickly relay messages for vital supplies, medical emergencies, and law enforcement. In addition to military purposes, the telegraph lines carried private and commercial messages across the territory, which brought in thousands of dollars to the US Treasury each year. (Carrie M. McLain Memorial Museum.)

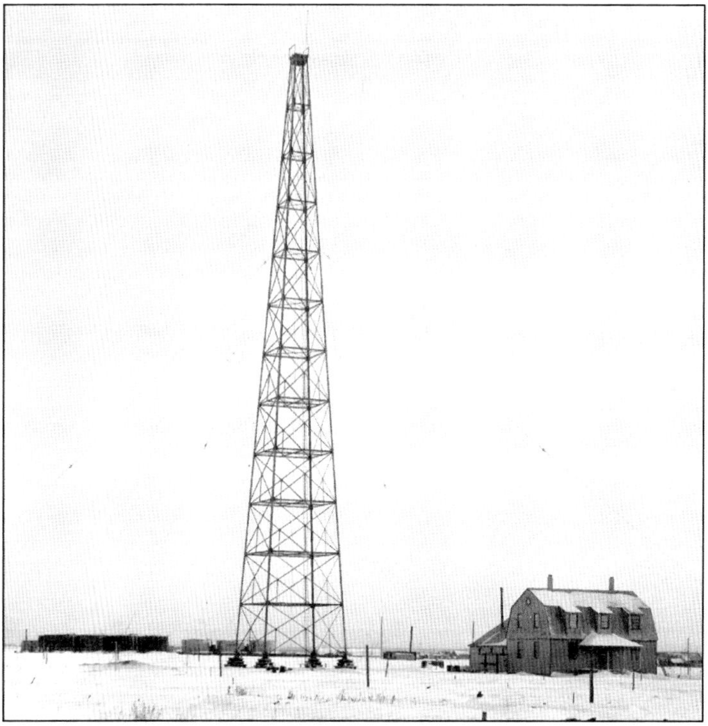

The US Wireless Telegraph Station at Nome was completed in the fall of 1908. The broadcasting tower soared 200 feet high and ran on 10 kilowatts of power, providing communication with ships several hundred miles out in the Bering Sea. (Photograph by F.H. Nowell; Carrie M. McLain Memorial Museum, McLain-322.)

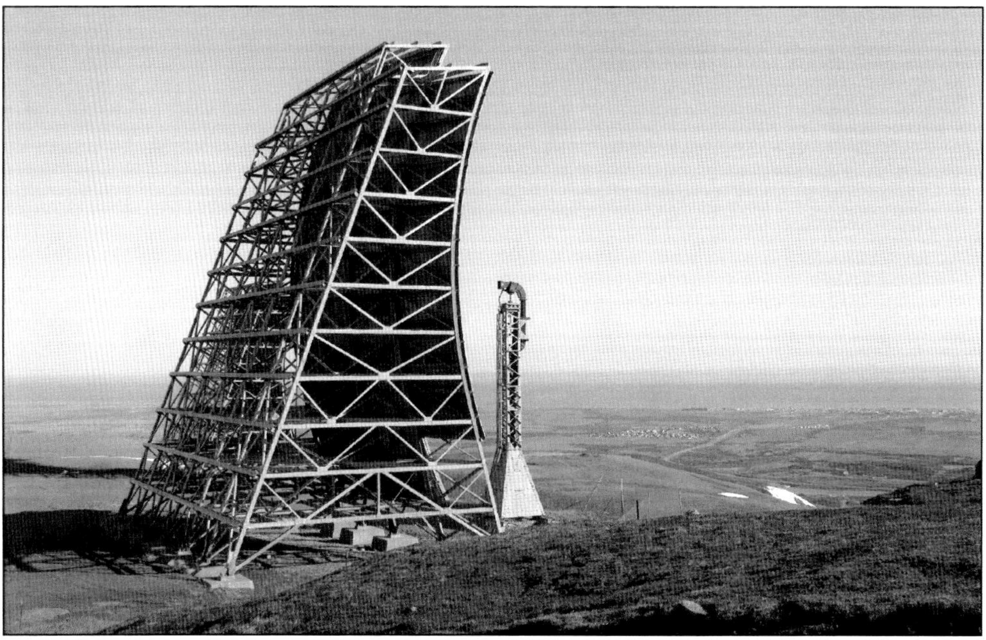

The US Air Force constructed the four large telecommunication panels of the Anvil Mountain White Alice Site in 1957. The panels were used during the Cold War to transmit signals between remote Air Force facilities in Alaska, such as the Distant Early Warning Line and Ballistic Missile Early Warning System. The system improved communication across rural Alaska and allowed Nome residents to place civilian phone calls. The steel framework of these giant antennae can still be seen on Anvil Mountain north of Nome. (Carrie M. McLain Memorial Museum.)

Nine

Dogs, Reindeer, and Horses

Draft animals, including dogs, reindeer, and horses, have played an important role on the Seward Peninsula for generations. Alaska Natives raised large and powerful huskies to pull sleds while hunting and traveling on the snow and ice. During the Nome gold rush, dog teams carried mail, hauled cargo from steamers anchored in the icy roadstead, and drew pupmobiles over railroad tracks to mining claims on the tundra.

In 1907, local dog team enthusiasts organized the Nome Kennel Club with a mission to advance the careful feeding and proper conditioning of sled dogs and to promote sled dog racing. From 1908 to 1917, the Nome Kennel Club sponsored the All Alaska Sweepstakes race, which ran 408 miles over sea ice, mountains, and tundra from Nome to Candle and back. Sled dog racing waned in Nome following the end of World War I but rose dramatically to the forefront again during the 1925 Serum Run, which brought a supply of life-saving diphtheria antitoxin to Nome. Since 1973, the arrival of March has encouraged professional mushers and their teams of 16 dogs to participate in the Iditarod Trail Sled Dog Race. The Iditarod covers 1,049 miles of rugged terrain from Willow to Nome, and mushers often speed through blizzards and sub-zero temperatures to complete the race in less than 10 days.

Reindeer arrived on the Seward Peninsula in 1892 as part of an effort by missionary and educator Sheldon Jackson to provide a reliable source of food, clothing, and transportation for Alaska Native communities. The first reindeer station was established at Port Clarence about 75 miles northwest of Nome, and herds were soon organized on the southern part of the Seward Peninsula. The reindeer population grew steadily from 10,000 in 1905 to more than 600,000 in 1934. Changes in ownership policy, range deterioration, predation, and disease greatly reduced the number of Alaska reindeer. The Kawerak Reindeer Herders Association currently oversees the Bering Strait population and a herd in the Nome area owned by the Midnite Sun Reindeer Ranch approximately 20 miles northeast of town.

Horses, mules, burros, and oxen were packed onto steamships for the journey to Nome in the early 1900s. Over 20 freighting companies operated in Nome at the height of the gold rush with horse-drawn wagons that hauled supplies across the tundra and into the mountains. Horse teams moved houses and dredges, carried water, and provided transportation for miners as well as soldiers stationed at Fort Davis. Mechanized farm tractors arrived in Nome during the 1930s and largely displaced the commercial use of dogs, horses, and oxen.

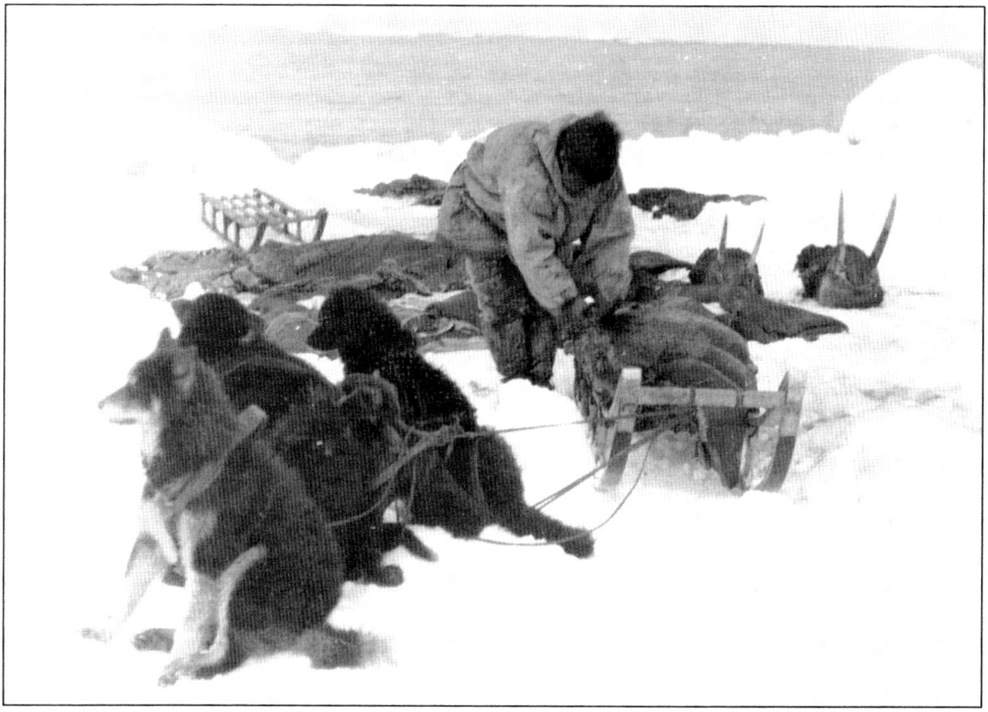

A husky team waits patiently as an Inupiat hunter loads a flat sled (*unirrat*) with walrus meat on top of the ice in the Bering Sea around 1910. Small pack sleds were typically used during short hunting trips while families hitched teams of five to nine dogs to longer railed sleds (*uniaġautit*) for winter travel. (Carrie M. McLain Memorial Museum, McLain-1378.)

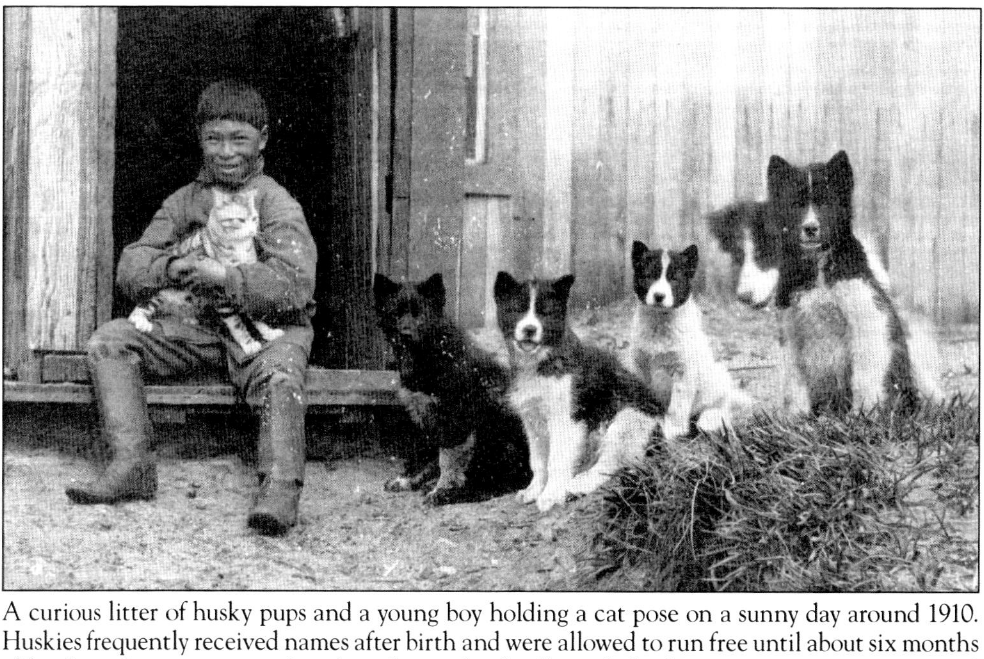

A curious litter of husky pups and a young boy holding a cat pose on a sunny day around 1910. Huskies frequently received names after birth and were allowed to run free until about six months old, when they were introduced to the work of pulling sleds. (Carrie M. McLain Memorial Museum, McLain-941.)

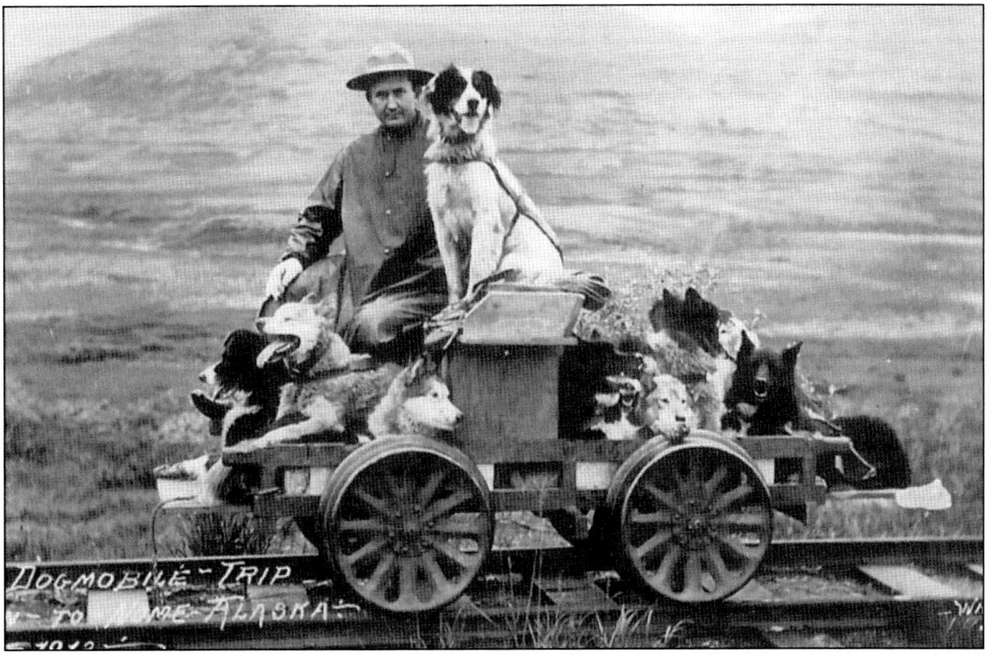

Enterprising miners used dog teams to pull small hand carts, called pupmobiles, over railroad tracks in the summer when tundra trails were thawed and muddy. Here, Walter W. Johnson, a gold dredge pioneer, and a team of sled dogs are traveling by pupmobile from Nome to Shelton on July 28, 1912. (Photograph by O.E. Wheeler; Carrie M. McLain Memorial Museum, Rolando-54.)

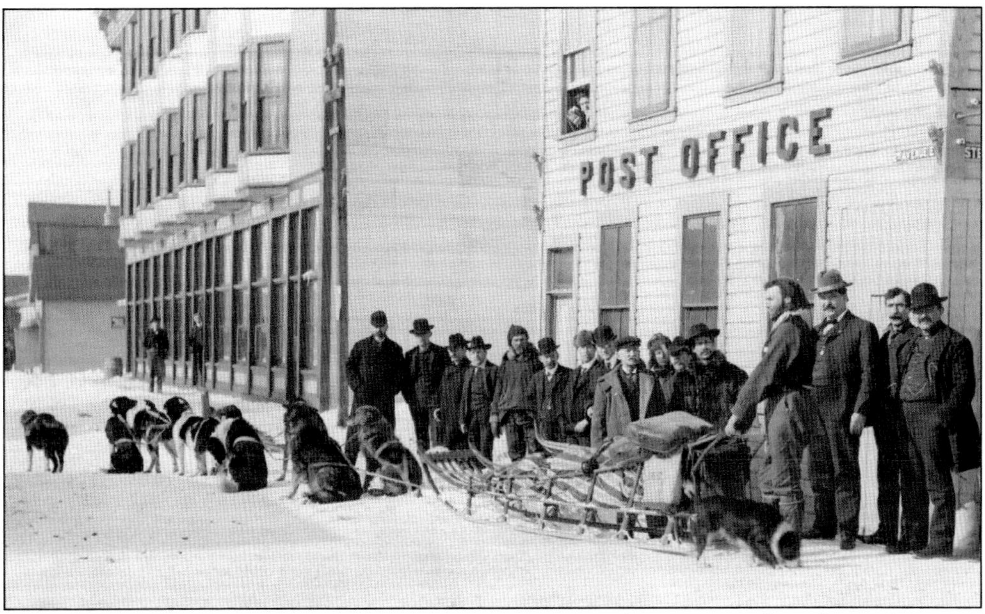

The winter arrival of a mail dog team after several months triggers a long line of anxious customers outside the post office at the corner of Second Avenue and Steadman Street in 1904. Mail carriers between Dawson and Nome covered hundreds of miles as they followed winter trails marked with stakes and branches on weekly mail routes. (Carrie M. McLain Memorial Museum, McLain-340.)

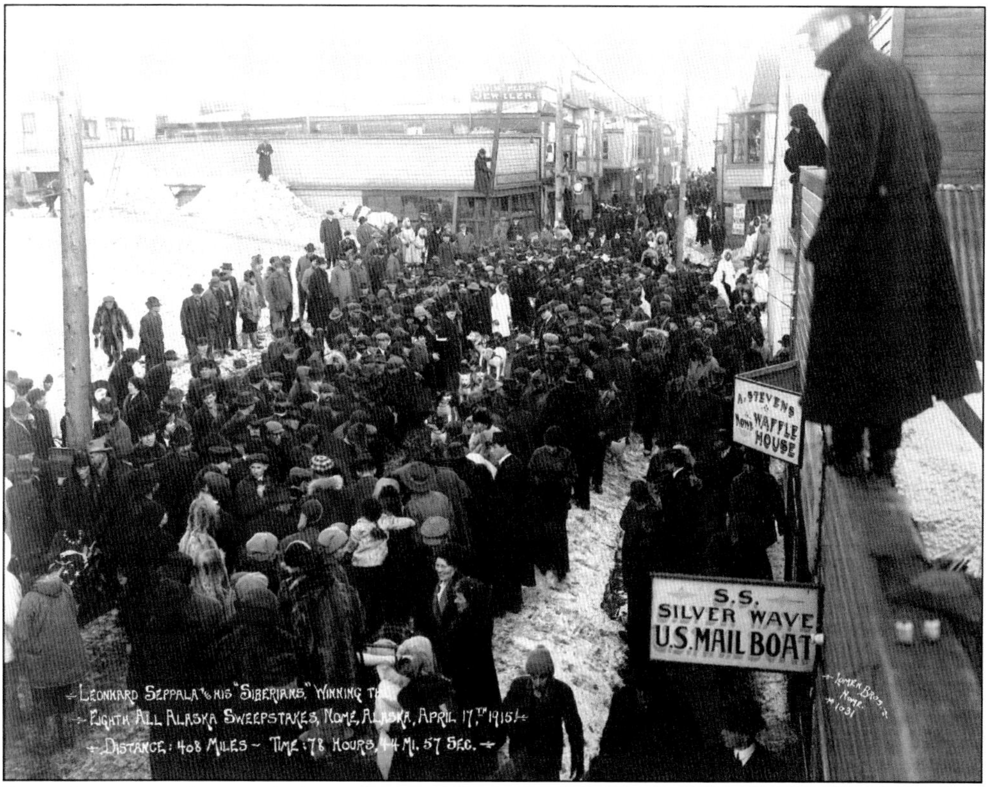

Front Street crowds struggle for a glimpse of first-place winner Leonhard Seppala and his team of Siberian huskies at the finish line of the eighth All Alaska Sweepstakes on April 17, 1915. Seppala trained and raced sled dogs from the kennel of the Pioneer Mining Company owned by Jafet Lindeberg. He consistently turned out impressive performances and won the All Alaska Sweepstakes three years in a row from 1915 to 1917. (Photograph by Lomen Brothers; Carrie M. McLain Memorial Museum, NMP-82-25-4.)

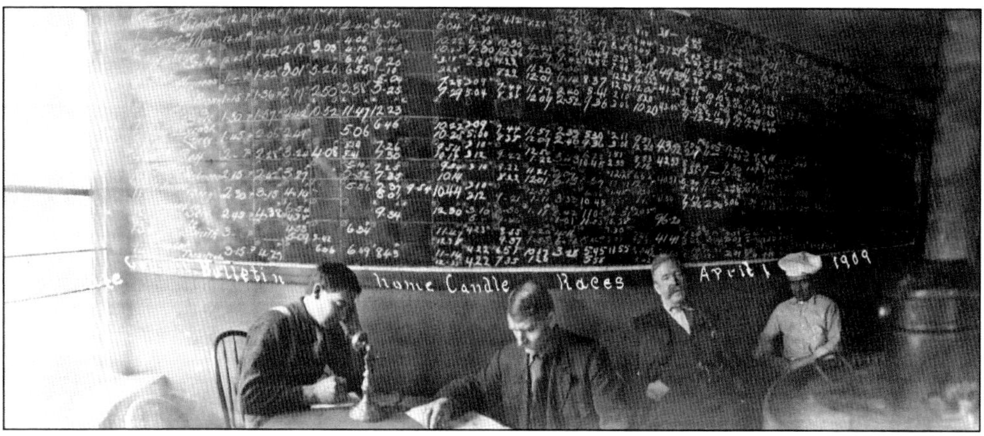

The Board of Trade saloon served as headquarters for the All Alaska Sweepstakes. There, race officials recorded reports of route times on a large blackboard, and eager spectators gambled on favorite teams. Fans treated drivers such as Leonhard Seppala, Scotty Allan, and Iron Man Johnson as celebrities with speaking engagements, film endorsements, and parades. (Carrie M. McLain Memorial Museum, McLain-1297.)

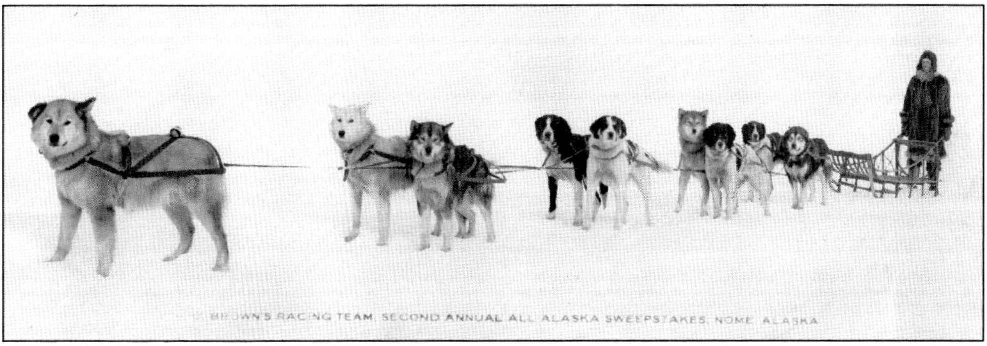

Driver H.O. Brown lines up his nine-dog team for an official photograph before the start of the second All Alaska Sweepstakes race in Nome in 1909. Sweepstakes drivers used light racing sleds and wore regulation clothing while dogs received nourishing meals of canned milk and raw ptarmigan. (Carrie M. McLain Memorial Museum, 2006.15.3.)

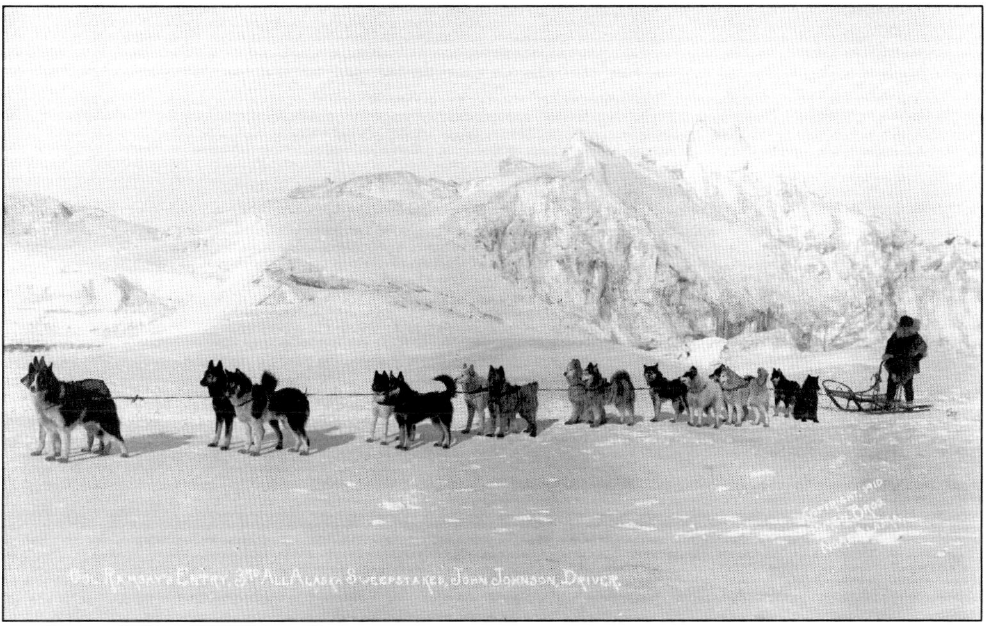

Iron Man John Johnson drove a 14-dog team of Siberian huskies owned by Col. Charles Ramsay into first place during the third All Alaska Sweepstakes in 1910. Johnson struggled through a blizzard and partial snow blindness to cover the 408 miles in just over 74 hours and win the $2,500 purse. (Photograph by Lomen Brothers; Carrie M. McLain Memorial Museum, 96.24.172.)

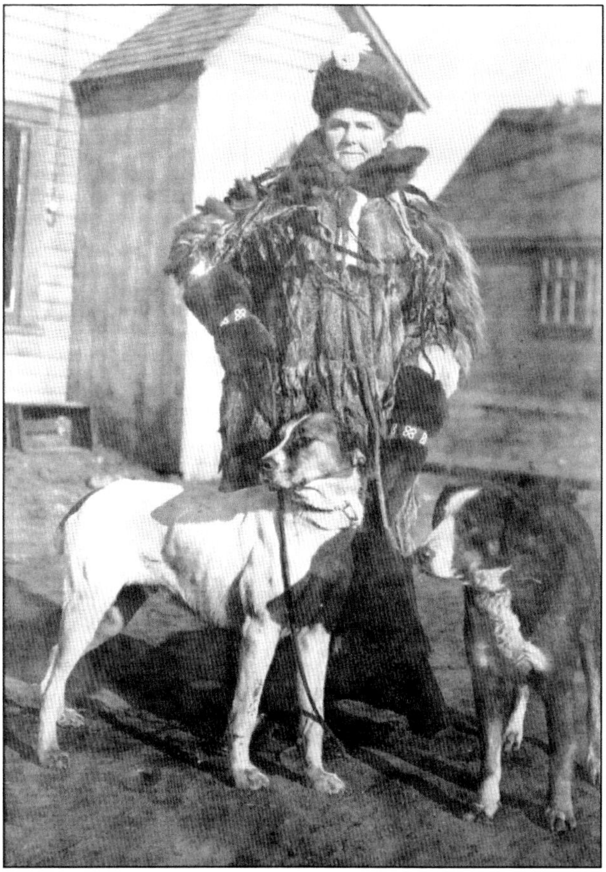

Esther Birdsall Darling poses in 1916 with two race dogs from the celebrated Allan-Darling kennel she co-owned with musher Scotty Allan. Darling helped found the Nome Kennel Club in 1907 and went on to become a successful author. Her popular book *Baldy of Nome* (1929) chronicles the true tale of a down-and-out dog who became a sled dog hero. (Carrie M. McLain Memorial Museum, McLain-1249.)

The Nome Kennel Club organized shorter speed races to encourage mushers to train during the winter and to foster interest in the All Alaska Sweepstakes in the spring. Interim races included the Solomon Derby, a loop from Nome to the Fort Davis Bridge, and an annual Ladies Race. Here, Barbara Bursik competes with five dogs in the 1916 Ladies Race as fans cheer her on along Front Street. (Photograph by Lomen Brothers; Carrie M. McLain Memorial Museum, McLain-1260.)

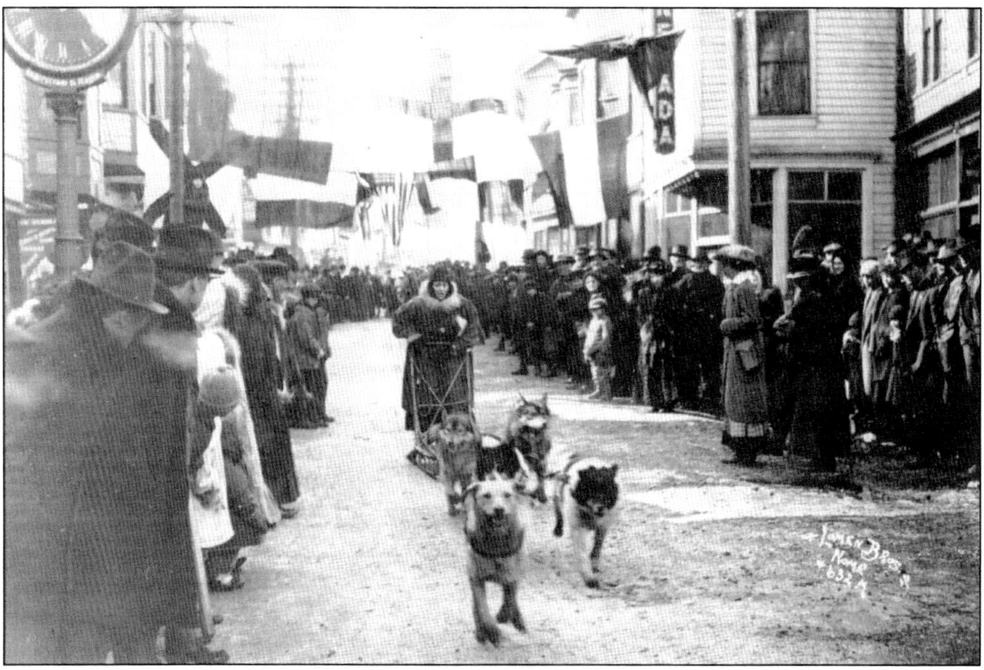

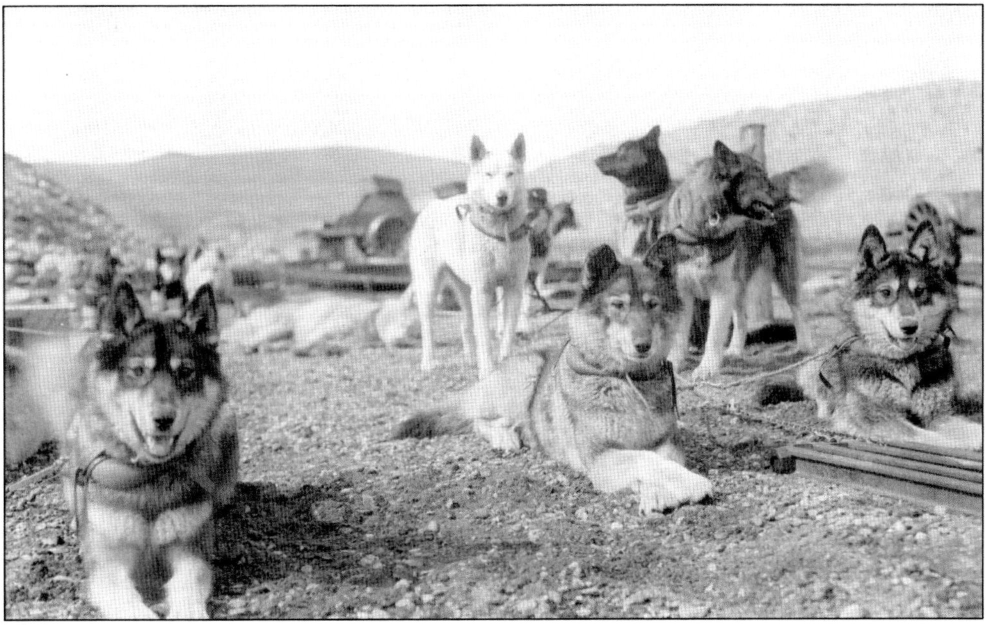

A sled dog team of Leonhard Seppala takes a break in their leather harnesses along the pubmobile track outside Nome around 1920. Famous lead dog Togo is in the center with his paws crossed. Togo led his teammates on several intrepid journeys including a 550 mile inspection through the interior of Alaska for the Alaska Railroad Commission in 1921, and a 260 mile leg of the relay to carry antitoxin to Nome during the 1925 Serum Run. (Carrie M. McLain Memorial Museum, 2015.7.)

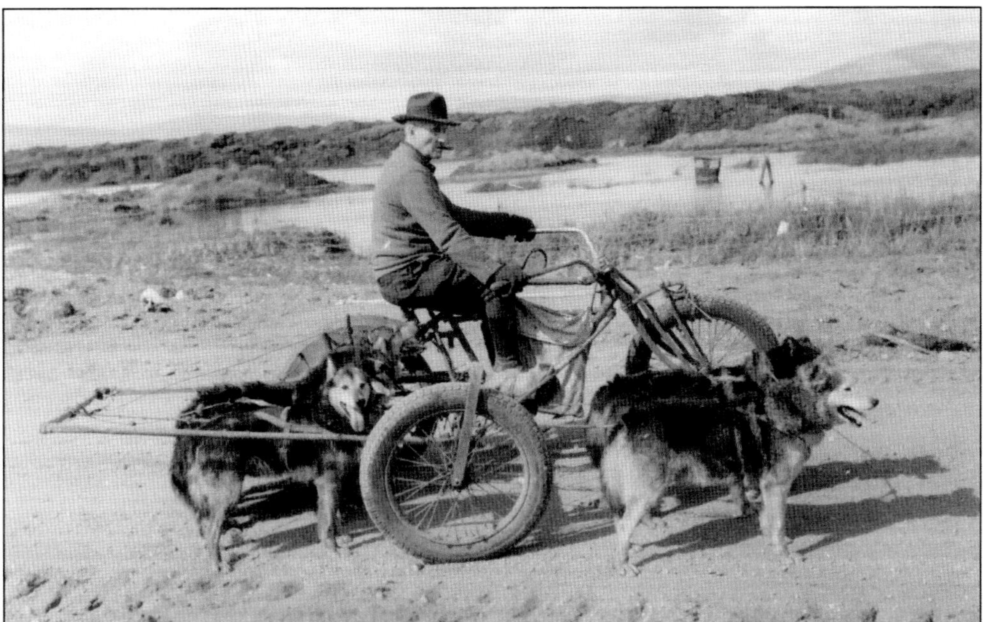

A modified bicycle with offroad tires provides exercise for a team of sled dogs in Nome between 1942 and 1943. Contemporary mushers use various techniques to train sled dogs during the offseason, including wheeled sleds, all-terrain vehicles, and electric treadmills. (Carrie M. McLain Memorial Museum, Potter-1529.)

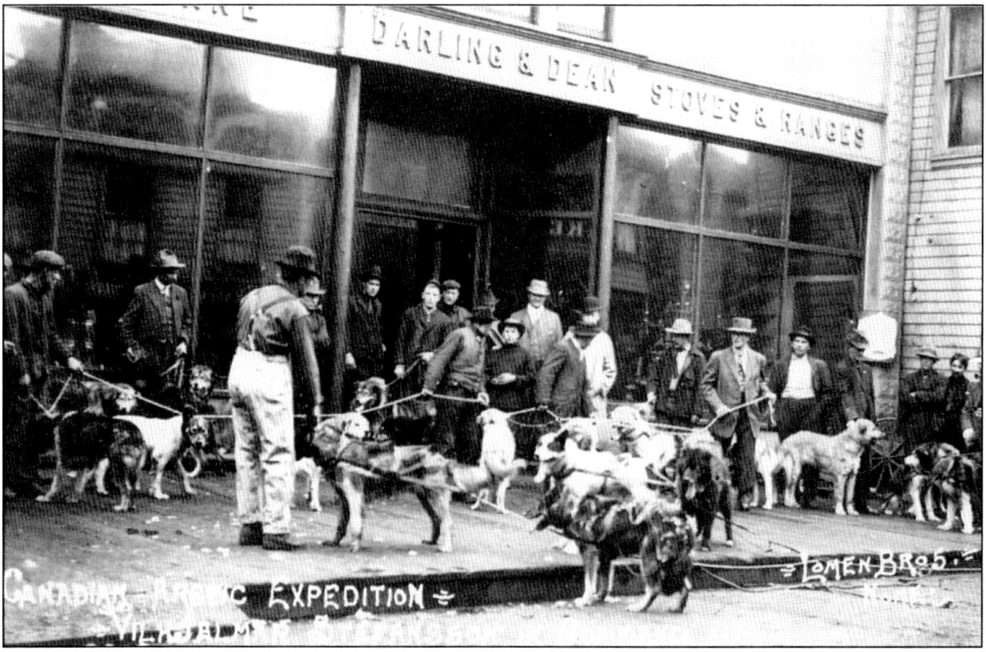

Arctic explorer and ethnologist Vilhjalmar Stefansson acquired supplies for the Canadian Arctic Expedition of 1913–1916 at the well-stocked Darling & Dean hardware store in Nome. The multidisciplinary expedition resulted in an immense collection of artifacts and photographs, film and sound recordings, and volumes of scientific data from the Arctic. (Photograph by Lomen Brothers; Carrie M. McLain Memorial Museum.)

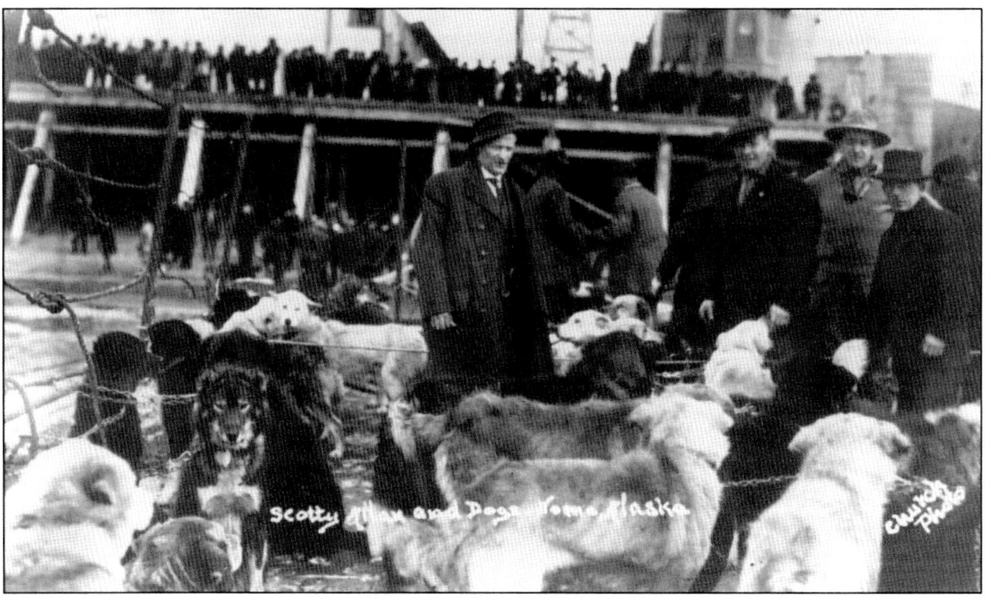

Scotty Allan organizes sled dogs on the Nome beach as they prepare to board a steamship for France in 1914. The US government commissioned 100 dogs from the Allan-Darling kennel to carry ammunition and transport wounded soldiers across the French Alps during World War I. Scotty Allan and Esther Birdsall Darling received the Croix de Guerre from the French government for their contributions to the war effort. (Carrie M. McLain Memorial Museum.)

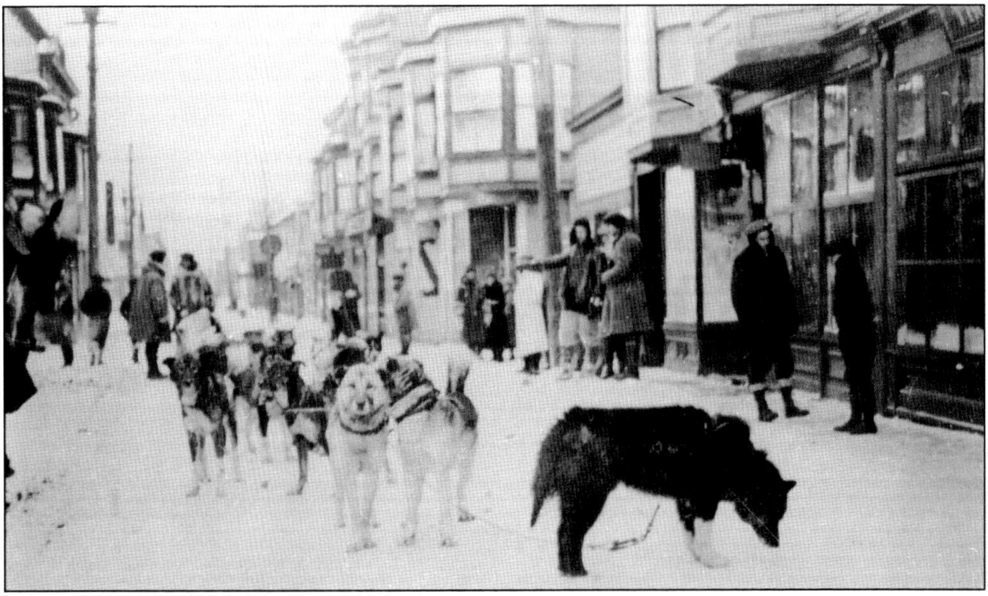

The 1925 Serum Run involved a relay team of 20 mushers and almost 150 sled dogs transporting vials of diphtheria antitoxin 674 miles from Nenana to Nome in five and a half days. The 13-dog team of musher Gunnar Kaasen is shown here on Front Street after their arrival on February 2, 1925, with celebrated sled dog Balto at the lead. (Carrie M. McLain Memorial Museum, NMP-85-23-349.)

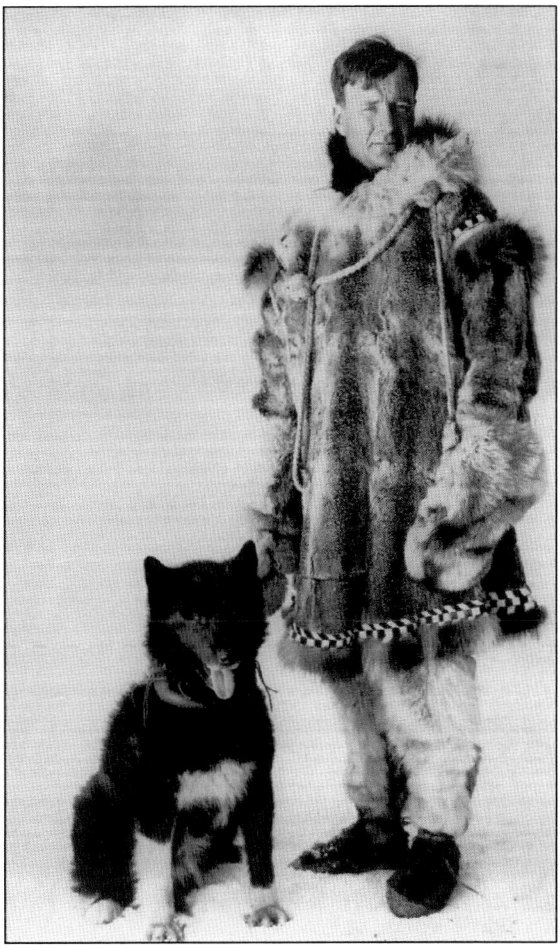

Gunnar Kaasen chose the large black-and-white freighting dog Balto from the kennel of Leonhard Seppala to lead his team and run the final leg of the 1925 Serum Run. The press hailed Kaasen and Balto as national heroes. New York erected a bronze sculpture of Balto in Central Park in 1925, and following his death in 1933, Balto was mounted for display at the Cleveland Museum of Natural History. (Carrie M. McLain Memorial Museum.)

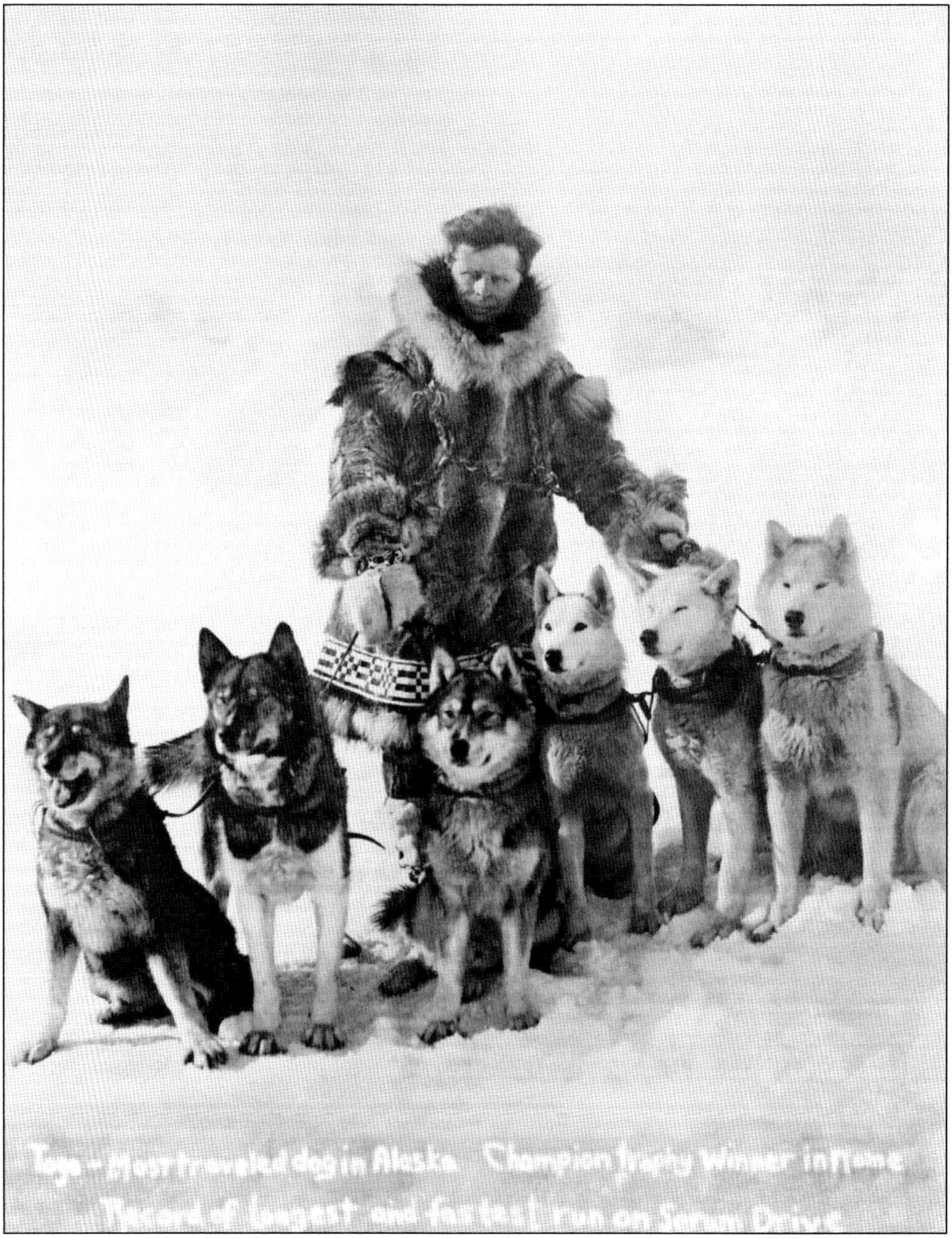

Leonhard Seppala stands with his prized dog team from the 1925 Serum Run led by Siberian huskies Togo (far left) and Fritz (far right). Seppala's team carried the package of serum on a 260 mile leg across jagged ice and through bone-chilling temperatures reaching -85°F. Fritz, a cream-colored husky bred by Leonhard Seppala, was a successful racing leader and an important foundation sire in early Siberian husky breeds. (Carrie M. McLain Memorial Museum, NMP-82-37-1.)

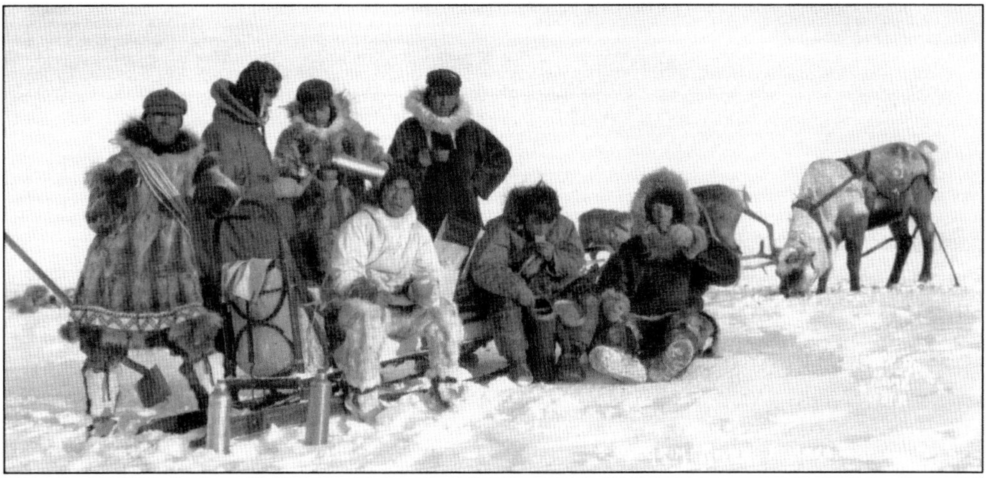

Travelers pause to enjoy mugs of hot drinks and biscuits while sled reindeer graze nearby on moss. From left to right are (seated) Marcus, Charlie, and Dr. Daniel Neuman; (standing) Toutuk, Walter C. Shields, Alfred, and Henry. Walter Shields served as the US Bureau of Education's superintendent of schools for the Northwest District of Alaska from 1910 to 1918. He advocated for increased schools throughout the region and expanded ownership of reindeer herds by Alaska Natives. (Carrie M. McLain Memorial Museum, 2017.5.)

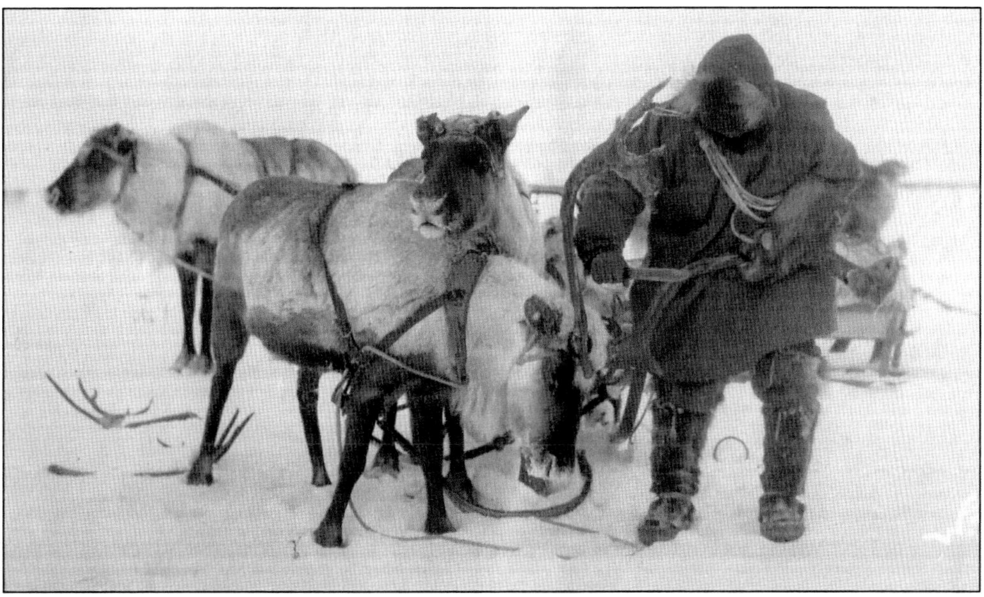

An Inupiat herder named Simon uses a knife to trim the horns of his reindeer while traveling between Selawik and Buckland in 1915. Approximately 1,300 reindeer were imported from Chukotka to Alaska between 1892 and 1902. Chukchi herders managed the first shipments of reindeer but soon returned to Russia. Saami herders from Lapland arrived in 1894 and trained local Alaska Natives in reindeer management, sled transportation, and animal husbandry. (Photograph by Walter Shields; Carrie M. McLain Memorial Museum, 2017.5.)

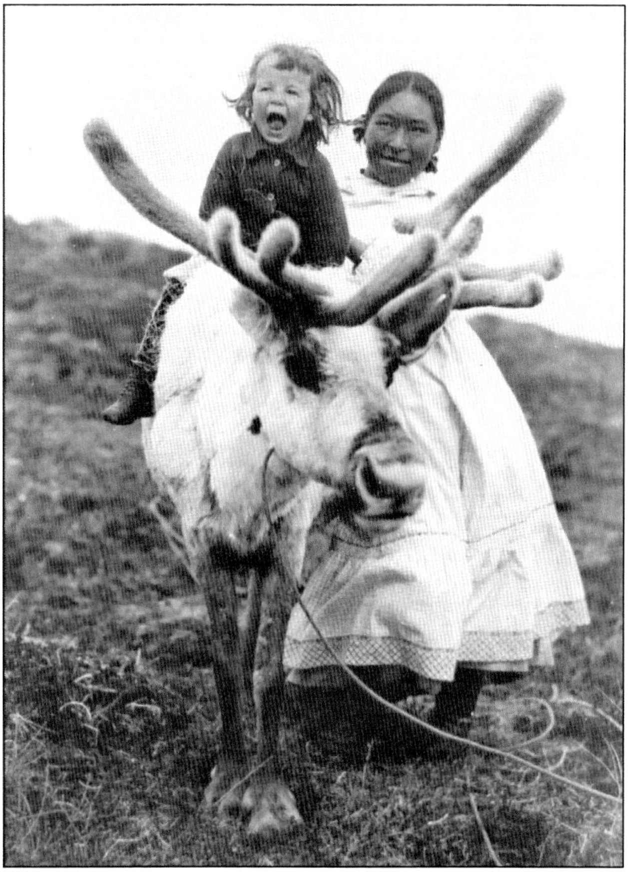

Sara Shields receives a supportive hand as she lets out an excited laugh from the back of a reindeer outside Nome on June 20, 1914. Semidomesticated reindeer are shorter, stockier cousins to wild Alaska caribou. Reindeer herds provide valuable meat, skins, antlers, milk, and transportation for Arctic communities. (Photograph by Walter Shields; Carrie M. McLain Memorial Museum, 2017.5.)

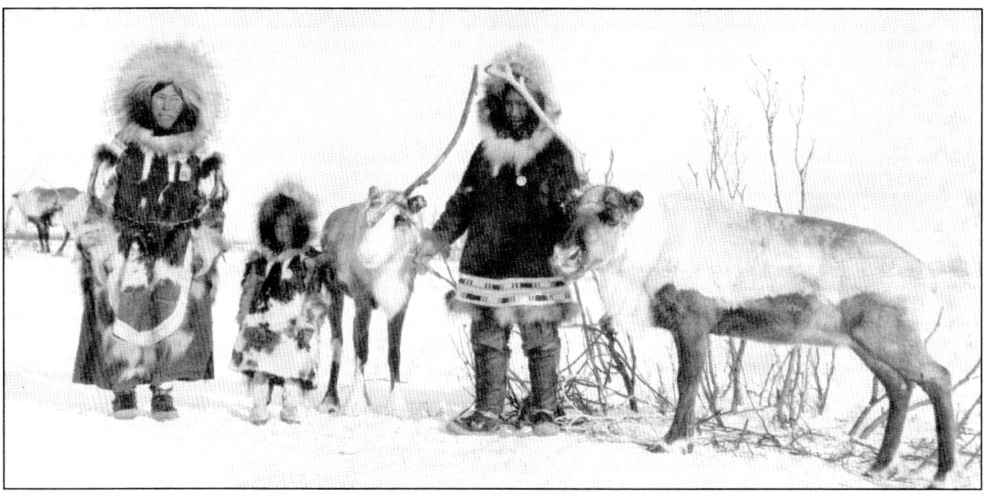

John and Mary Oquilluk pose with their son William during the reindeer fair at Mary's Igloo in 1915. The US Bureau of Education organized annual reindeer fairs from 1915 to 1918 to increase Alaska Native interest and skill in the reindeer herding industry. Young William Oquilluk went on to record Inupiat oral traditions and stories that he compiled and published in 1973 as *People of Kauwerak: Legends of the Northern Eskimo.* (Photograph by Walter Shields; Carrie M. McLain Memorial Museum, 2017.5.)

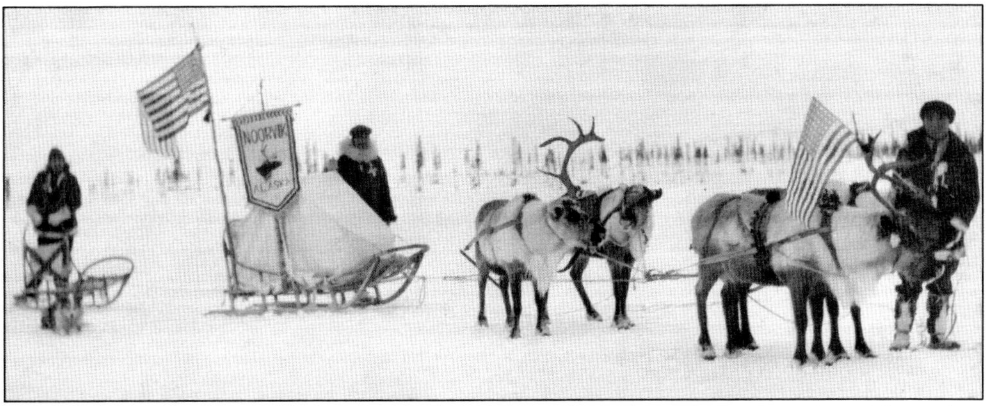

Flags and banners wave in the winter breeze as herders from Noorvik arrive at the 1915 reindeer fair at Mary's Igloo. The fair offered herders an opportunity to show off their skills through harness and sled shows, shooting matches, and deer races. (Photograph by Walter Shields; Carrie M. McLain Memorial Museum, 2017.5.)

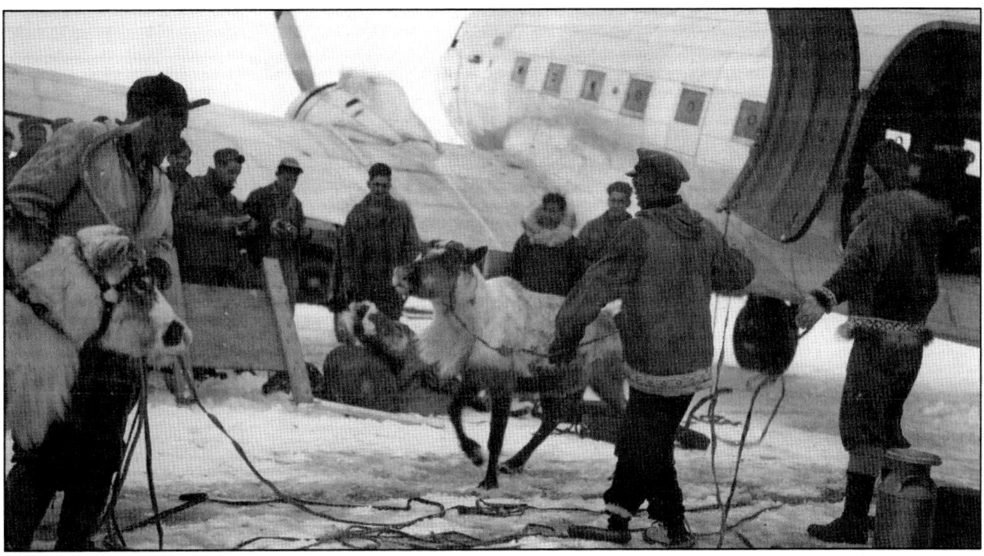

The US Air Force offered its assistance and airlifted reindeer from Kotzebue to Nome for a reindeer fair organized by the US Reindeer Service in 1947. Chester Asakak Seveck served as the government appointed manager of the Kotzebue, Selawik, and Buckland reindeer herds during the 1940s and documented his experiences in the 1973 book *Longest Reindeer Herder*. (Carrie M. McLain Memorial Museum, 2006.22.1914.)

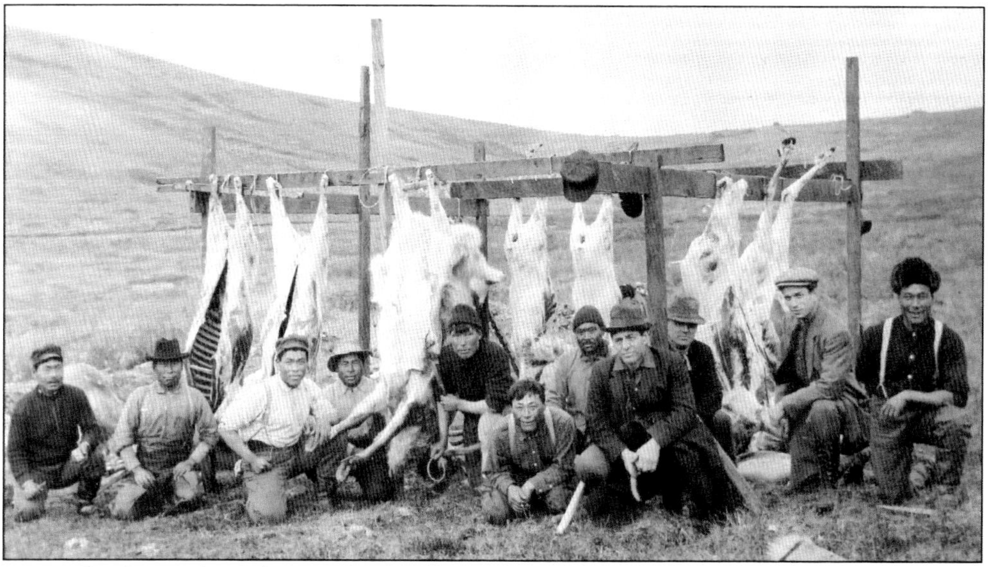

Inupiat herders and reindeer station superintendent William T. Lopp (fourth from right) from Wales pose in front of reindeer they brought to Nome to be to be butchered and sold in 1914. Reindeer were generally field processed during the month of October when the snow was clean and subzero temperatures would freeze the carcasses. (Photograph by Walter Shields; Carrie M. McLain Memorial Museum, 2017.5.)

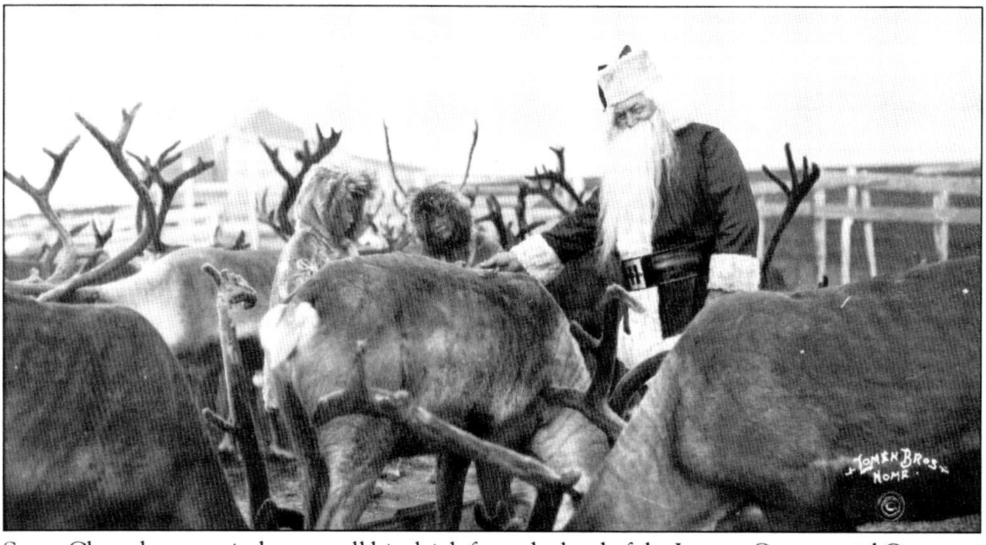

Santa Claus chooses reindeer to pull his sleigh from the herd of the Lomen Commercial Company in Nome around 1928. The Lomens significantly advanced the promotion and development of the Alaska reindeer industry during the late 1920s. The company took reindeer on tours in the Lower 48, produced cookbooks, and built corrals, slaughterhouses, and refrigerated ships to transport reindeer meat. (Carrie M. McLain Memorial Museum.)

A mule catches a bird's-eye view of Nome as he is swung off a lighter and onto the beach around 1900. Horses and mules were in high demand during gold rush Nome as they provided critical transportation of people and supplies across the tundra. (Carrie M. McLain Memorial Museum, McLain-858.)

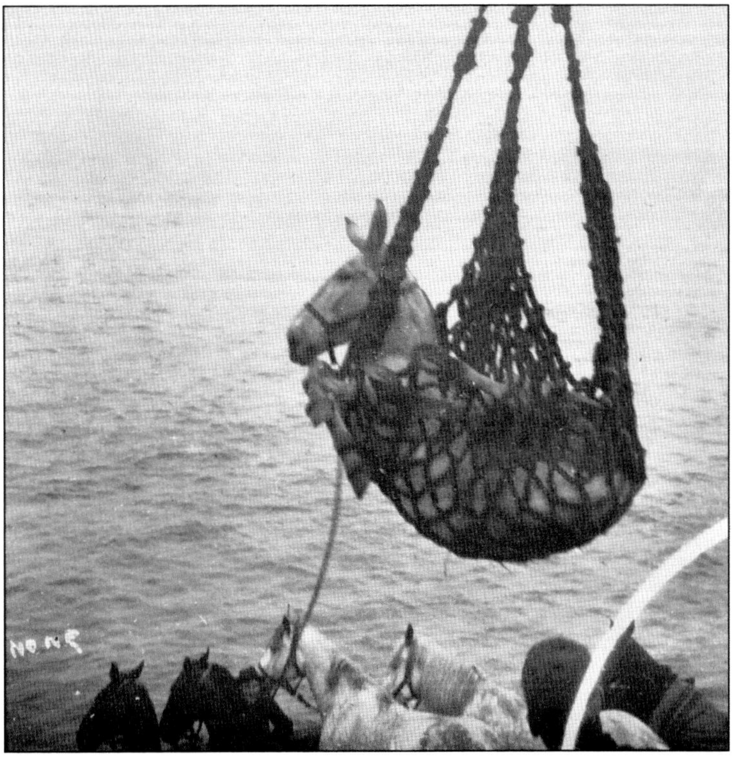

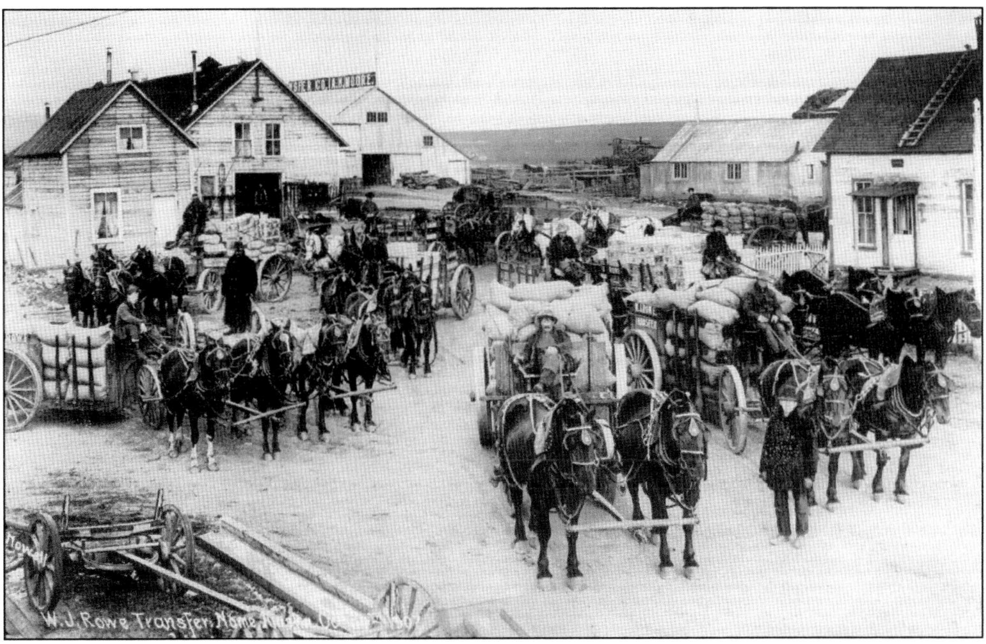

Teams of horses from the W.J. Rowe Transfer Company prepare to convey heavily laden wagons of supplies to mining outfits across the Seward Peninsula in 1907. Billy Rowe began his company with a horse he drove from Dawson over the frozen Yukon and into Nome. His freighting outfit grew to one of the largest in Nome with over 100 draft horses. (Photograph by F.H. Nowell; Carrie M. McLain Memorial Museum, 2011.11.)

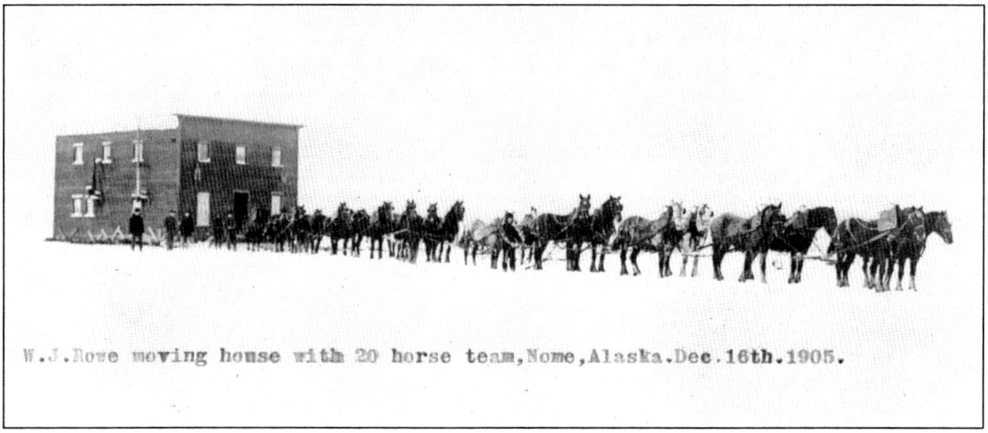

A team of 20 horses belonging to the W.J. Rowe Transfer Company prepare to move a house on runners across the snow in Nome on December 6, 1905. Moving buildings into and out of town was a common recycling practice due to the high cost of shipping in construction materials. (Carrie M. McLain Memorial Museum, 2011.11.)

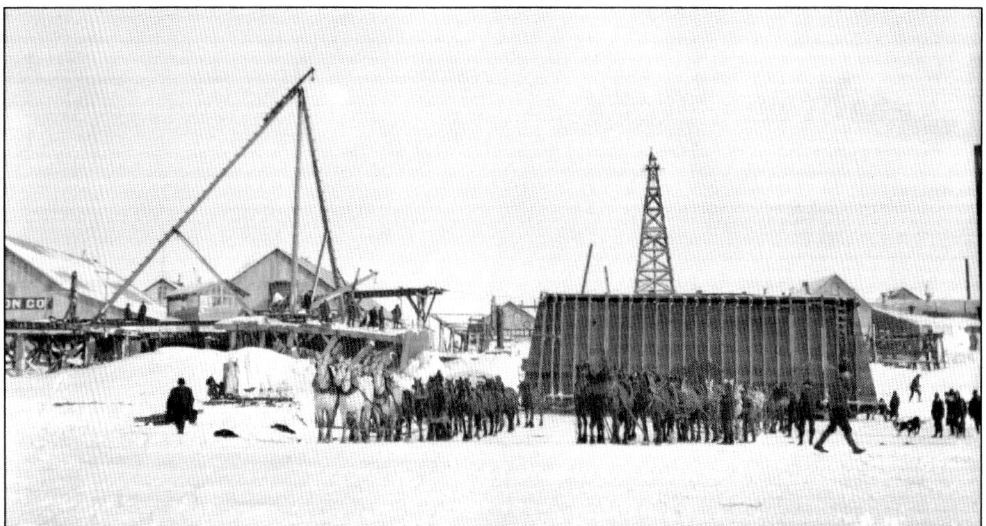

Sixty-four horses are harnessed to an immense floating dredge with a caisson drill as they prepare to move it out onto the frozen Bering Sea to mine for gold. Caisson drills are capable of creating large-diameter holes through thick layers of ice, solid bedrock, or alluvial gravels. (Photograph by Arthur L. Bell; Carrie M. McLain Memorial Museum, Bell-N-248.)

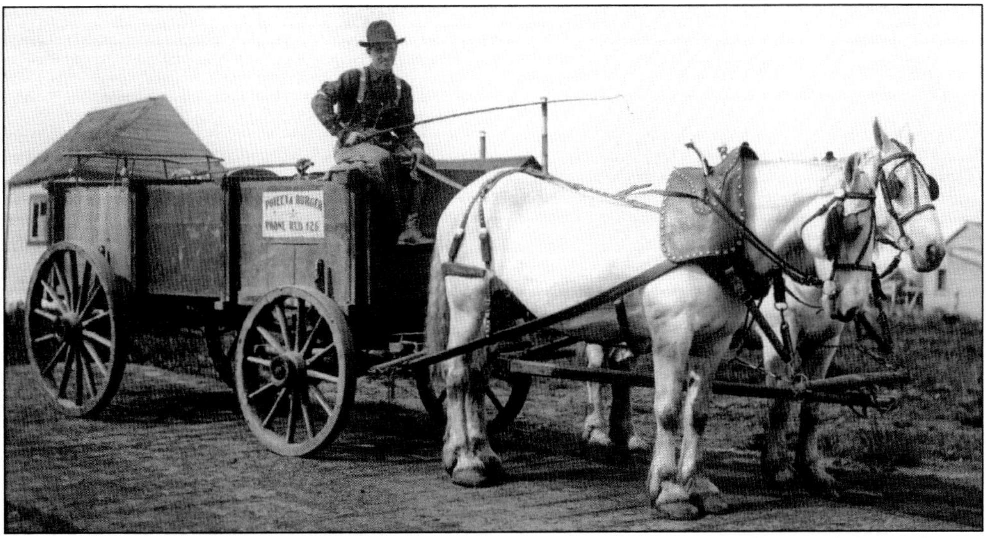

A horse-drawn wagon operated by the Poteet and Burger company offered water delivery during the winter months when Nome residents were unable to access water from the frozen mains stretching four and a half miles to Moonlight Springs. A $1 ticket could typically purchase five gallons of water from a water wagon. (Carrie M. McLain Memorial Museum, McLain-344.)

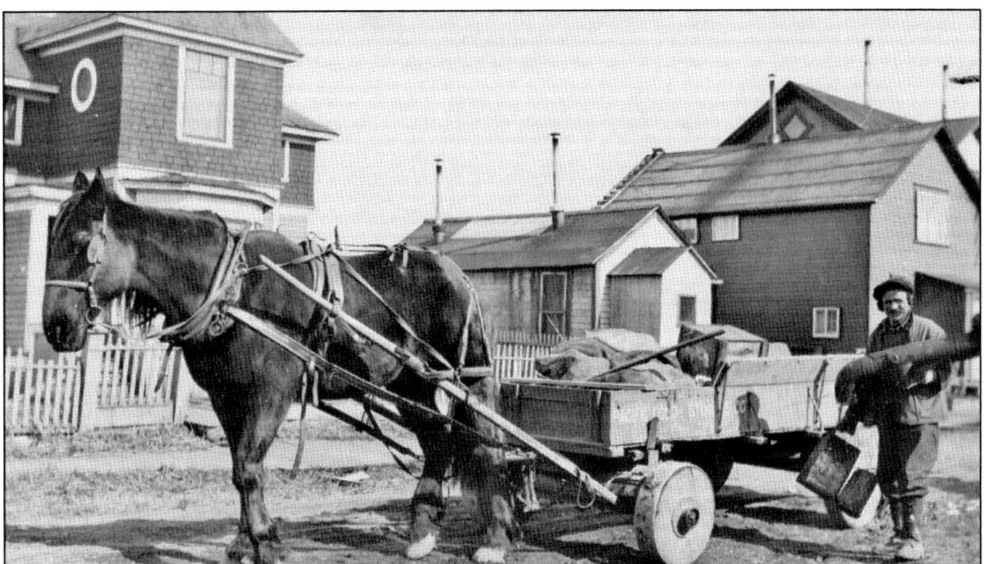

Eli Nicolai, shown here between 1905 and 1910, traveled with his horse and wagon around town to gather castoff items and provide a honey bucket service before the installation of indoor lavatories in Nome. His wagon is parked in front of the Jacob Bergher House (far left), which was built with Queen Anne–style details in 1904 and listed in the National Register of Historic Places in 1977. (Carrie M. McLain Memorial Museum, McLain-249.)

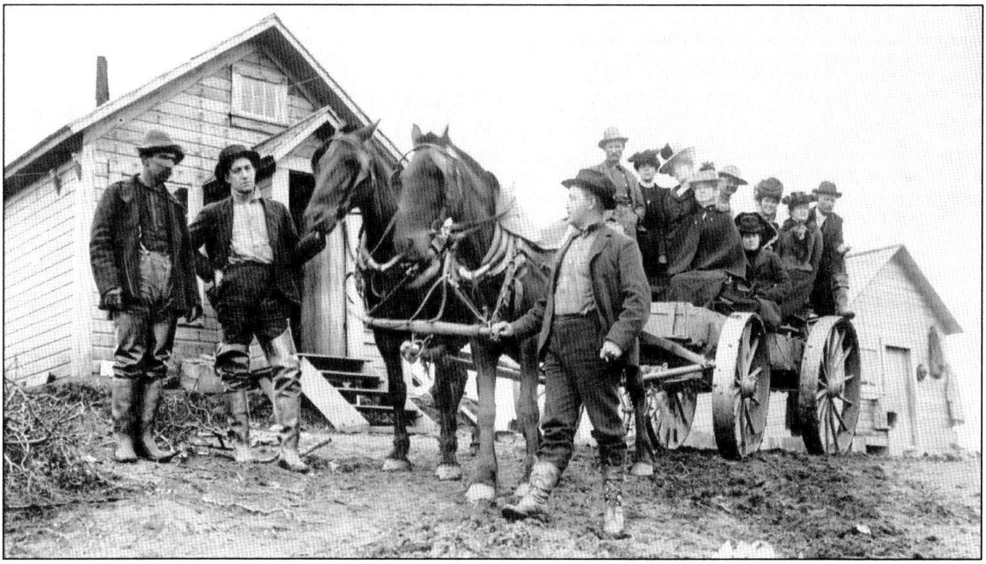

Nome residents crowd onto a horse-drawn wagon after touring the Hot Air Bench Mine, located above Glacier Creek north of Nome, around 1901. Excursions to large mining operations proved a popular form of educational entertainment for the local townsfolk. (Photograph by B.B. Dobbs; Carrie M. McLain Memorial Museum, 96.26.17.)

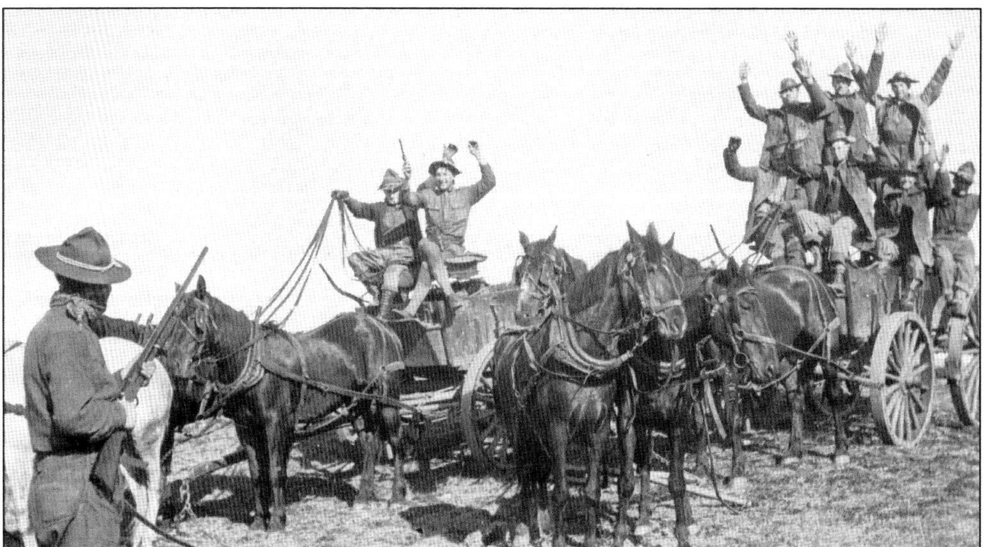

Horse-drawn wagons carry soldiers from Fort Davis, who share a lighthearted moment as they raise their arms in mock surrender around 1913. Fort Davis was established a few miles east of Nome in 1900, and soldiers were tasked with helping to maintain order and improve sanitation within the growing community. (Carrie M. McLain Memorial Museum, 1997.6.)

Ten

Wings and War

Nome residents looked to aircraft to bridge the distance between their isolated community and the outside almost as soon as the dust had settled from the gold rush. The first plane in Nome was built in 1911 by local music teacher Henry Peterson and came equipped with skis, a gas engine, and a push propeller. The plane moved along the ground for several feet but failed to take flight. The 1920s were an era of innovation. The Black Wolf Squadron flew from New York to Nome and confirmed that aviation in Alaska was possible, Noel Wien flew the first commercial roundtrip flight from Fairbanks to Nome, and Norwegian explorer Roald Amundsen and the crew of the dirigible *Norge* made the first transpolar flight from Europe to North America.

Alaska visionaries of the 1930s argued that Nome could become an international aviation base due to its strategic coastal location. Although never a center for worldwide air traffic, Nome did develop into a central base for Alaskan air carriers. By 1939, five major airlines operated year-round in Nome including Wien Alaska Airlines, Mirow Air Service, Ferguson Airways, Northern Cross, and Pacific Alaska Airways.

At the outbreak of World War II in 1939, the only military base in Alaska was an outdated post in the southeast with a few hundred troops. Congress decided Nome would be a strategic location for a bomber base that could enable long-range patrols of the Bering Sea and protection of Alaska's west coast. Construction of Marks Air Force Base began in 1941; upon completion, it served as a garrison for hundreds of men, aircraft, and military supplies. The base was reconfigured into a municipal airport for Nome following its decommission in 1956.

Air travel to and from Nome is now a habitual part of life for the majority of residents. Nome serves as a hub for passenger and freight service in northwest Alaska. Commercial giant Alaska Airlines offers direct service between Nome and Anchorage, and Bering Air offers scheduled service to 32 remote Alaska villages and international service to Russia. In 2007, the air freight company Northern Air Cargo began operating Boeing 737-200 aircraft with a capacity to transport 30,000 pounds of freight to Nome.

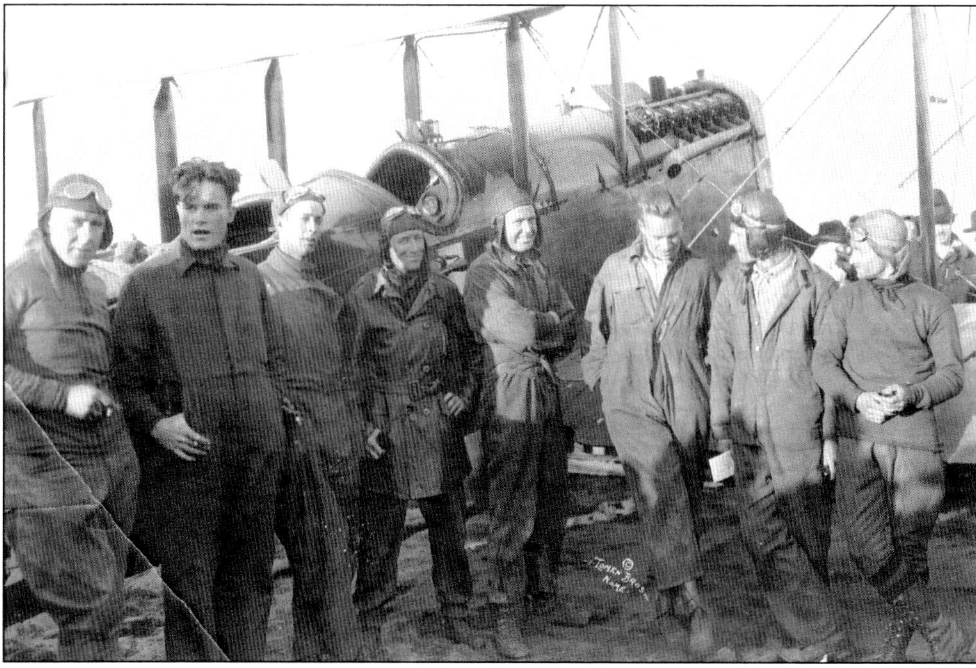

The US Army sent the eight-member Black Wolf Squadron on a monthlong publicity tour in July 1920 from New York to Nome in four de Havilland DH-4 biplanes. Pilots logged 50 hours of flight time on their journey to Nome and proved that aviation in Alaska was possible. (Photograph by Lomen Brothers; Carrie M. McLain Memorial Museum, McLain-1221.)

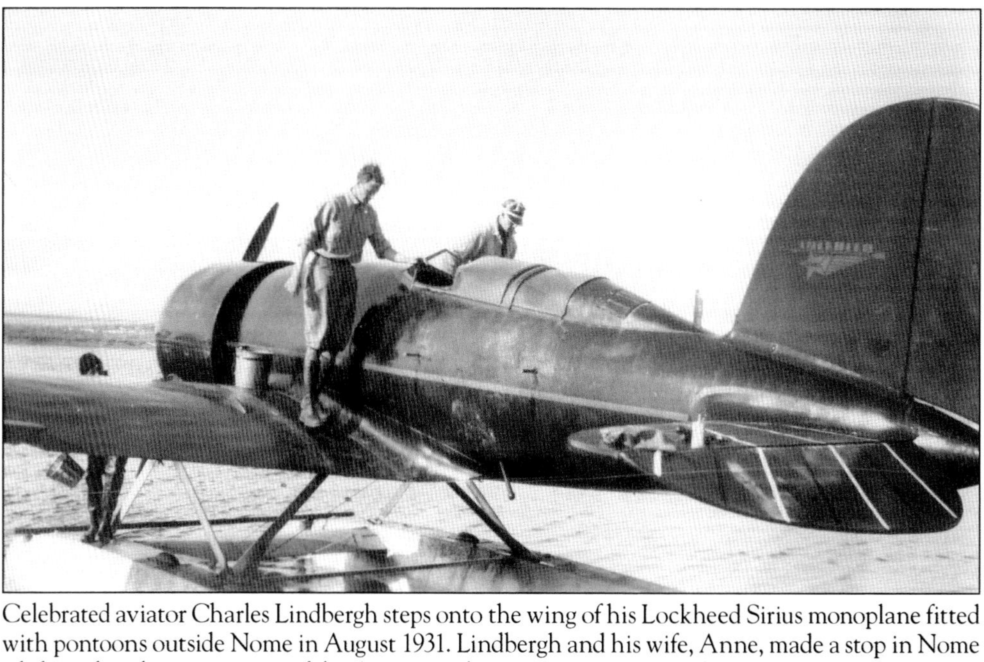

Celebrated aviator Charles Lindbergh steps onto the wing of his Lockheed Sirius monoplane fitted with pontoons outside Nome in August 1931. Lindbergh and his wife, Anne, made a stop in Nome while undertaking a journey of the Arctic Circle to scout a commercial air route for Pan American Airways. (Photograph by Lomen Brothers; Carrie M. McLain Memorial Museum, Jacobs-879.)

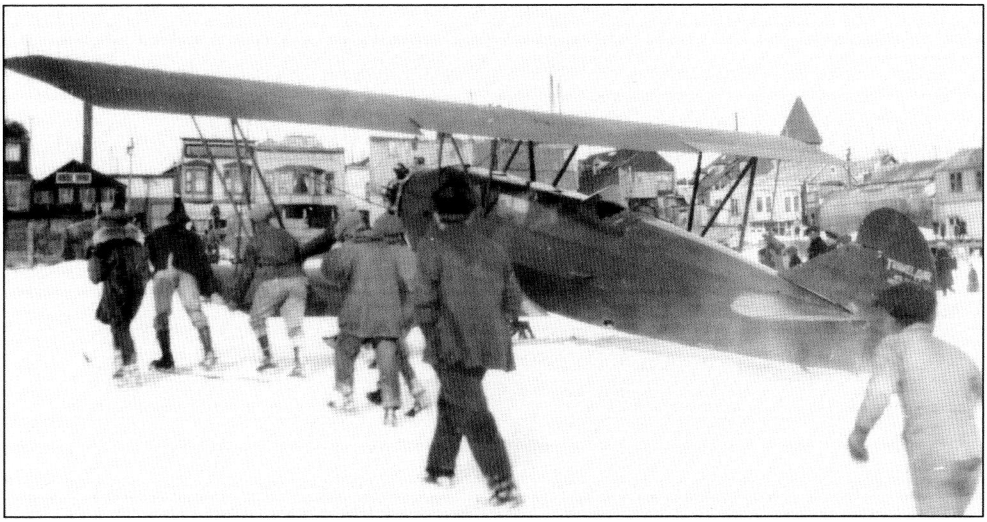

A Travel Air biplane carrying pilot R.A. Merrill and physician A. Cope lands on the ice outside Nome on April 4, 1927. The pair undertook a medical mission from Anchorage to Savoonga on St. Lawrence Island, but due to the length of the flight, did not reach the village in time. Their unfortunate experience helped propel the formation of an airline service based in Nome. (Carrie M. McLain Memorial Museum, NMP-83-1-122.)

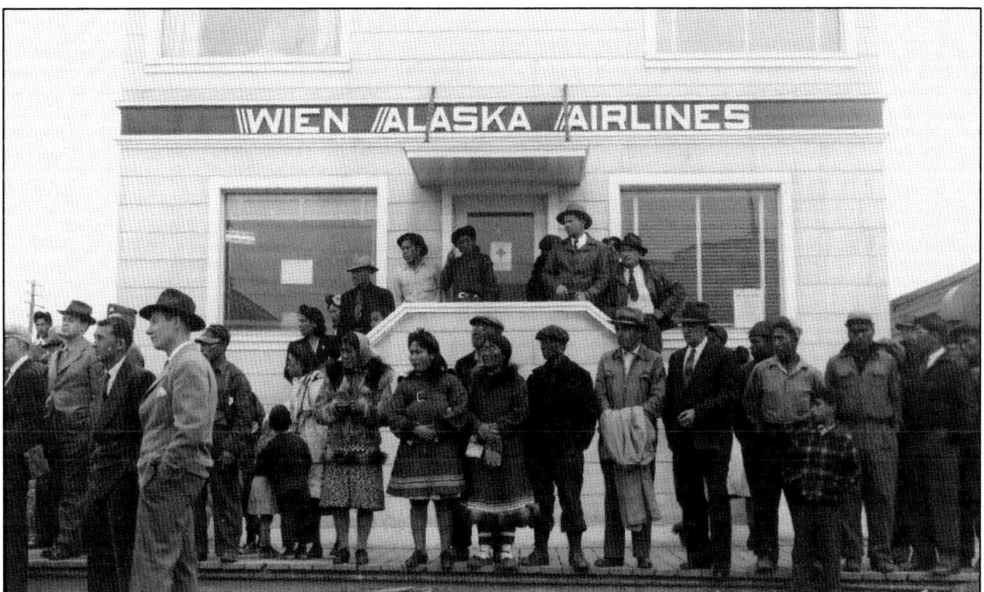

A crowd gathers to watch a parade in front of the offices of Wien Alaska Airlines in Nome during the 1940s. Experienced bush pilot Noel Wien offered the first scheduled passenger flights from Nome to Fairbanks in June 1927. The family business expanded to provide airplane and helicopter service throughout Alaska and into the Western United States until the 1980s. (Carrie M. McLain Memorial Museum, Johnson-1357.)

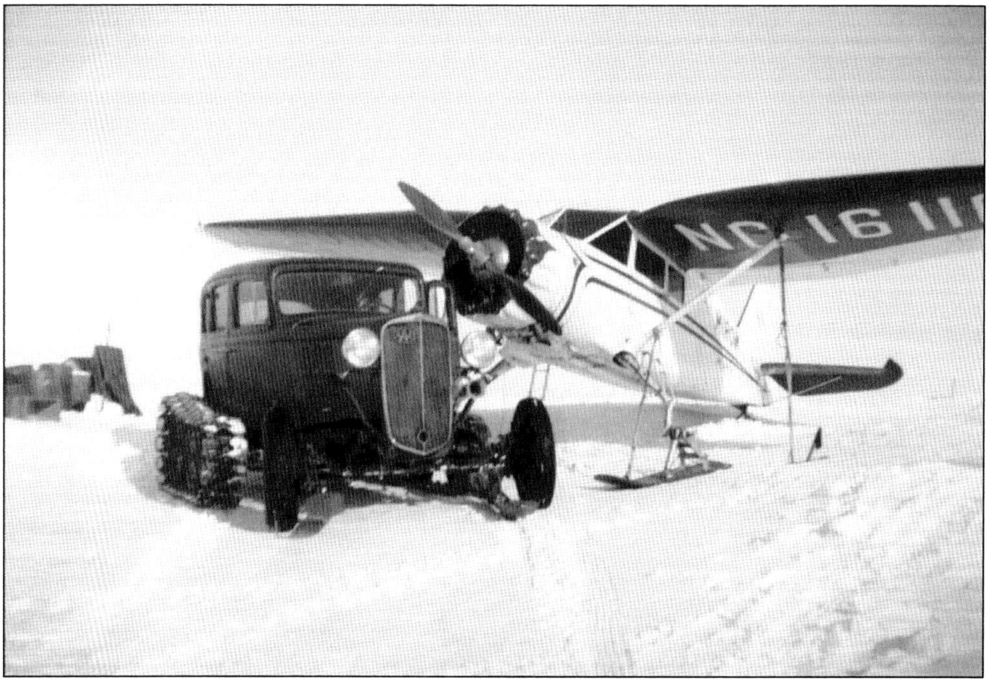

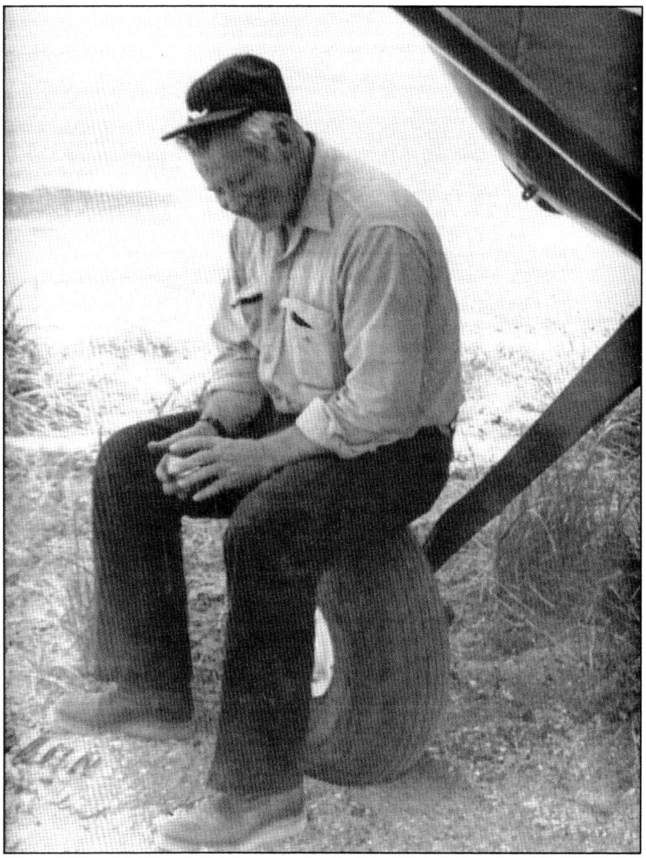

Hans Mirow formed Mirow Air Service in Nome in 1935. The air carrier offered scheduled deliveries of mail and passengers with tracked taxi service available during the winter for customer convenience. (Carrie M. McLain Memorial Museum, NMP-97-16-20.)

Bush pilot Neal "Willie" Foster rests on the tire of a plane during a sunny day in Nome. Willie Foster started Foster Aviation in 1946 after returning to Nome from World War II. His son Richard Foster took over management of the air service following Richard's return from the Vietnam War. The company carried a reputation as the "Friendliest Air Carrier" and was well known for its Cessna 185 airplane, known as "9-5 Charlie." (Carrie M. McLain Memorial Museum.)

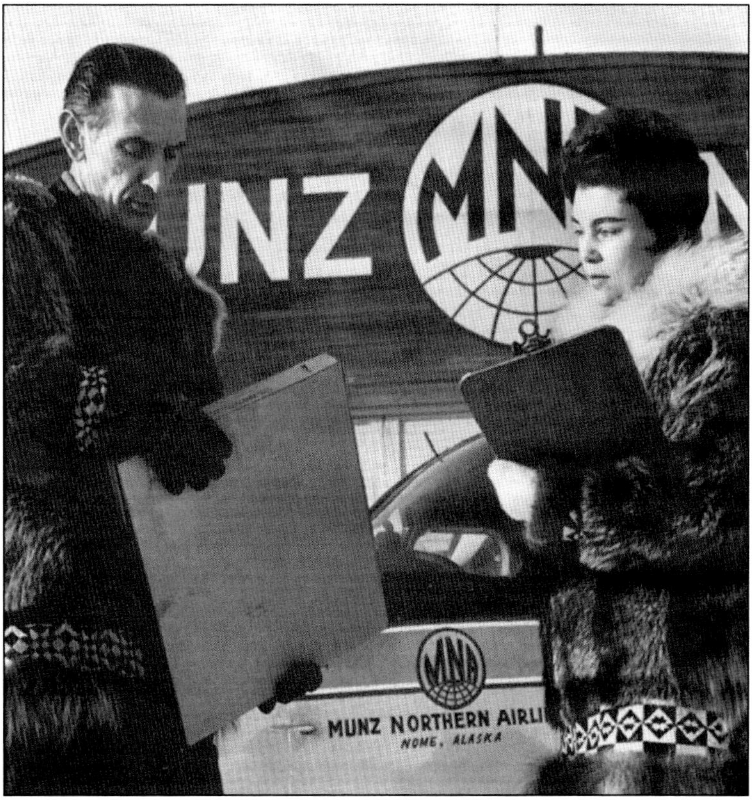

In 1962, Dick and Joyce Galleher purchased Munz Airways from Bill Munz, who started the company in Nome in 1938. The couple renamed the air service Munz Northern Airlines and began a 20-year career of providing transportation to over 50 villages in Alaska from Point Hope to Bethel. (Photograph by Bob and Ira Spring; Carrie M. McLain Memorial Museum.)

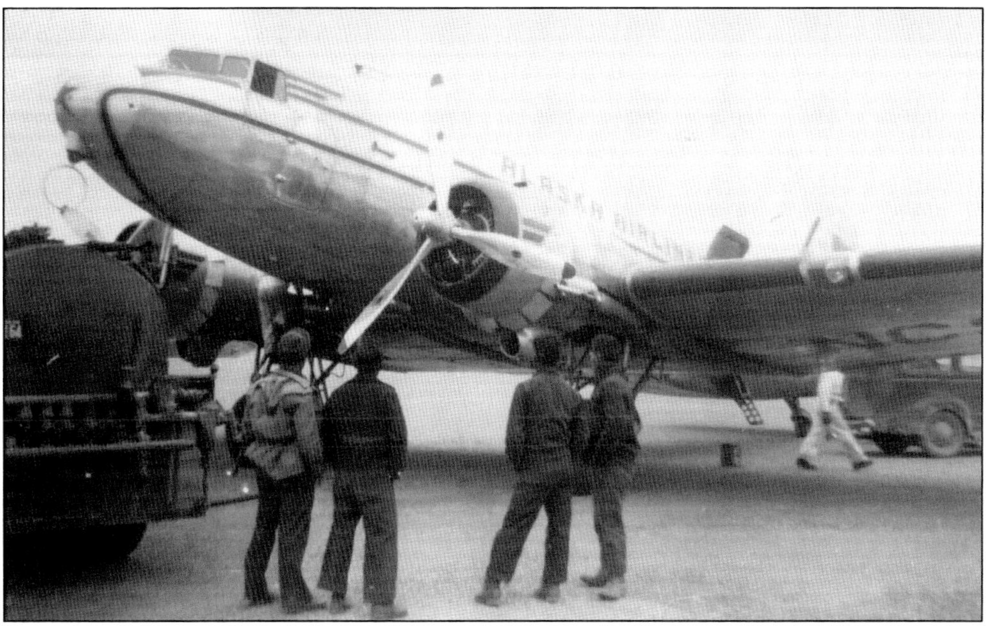

Alaska Airlines traces its roots to McGee Airways, which began offering service in 1932. The company greatly expanded during the 1940s and added surplus World War II military aircraft to its fleet, such as the Douglas DC-3 shown here in Nome. (Carrie M. McLain Memorial Museum, Johnson-1099.)

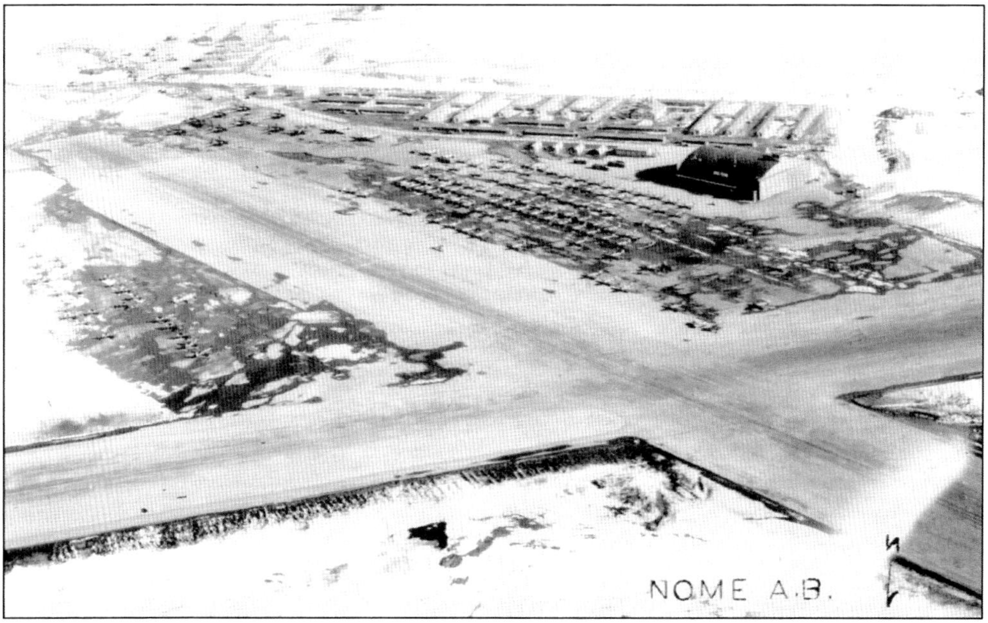

Marks Air Force Base was built across the Snake River on tailing piles left behind by gold dredges. The tailings provided a firm foundation for buildings, roads, and landing strips. Construction of the base began in July 1941, and by the end of 1942, it included 776 buildings and a landing field with two runways that could handle any type of airplane the United States had in production at the time. (Carrie M. McLain Memorial Museum, 2003.35.2190.)

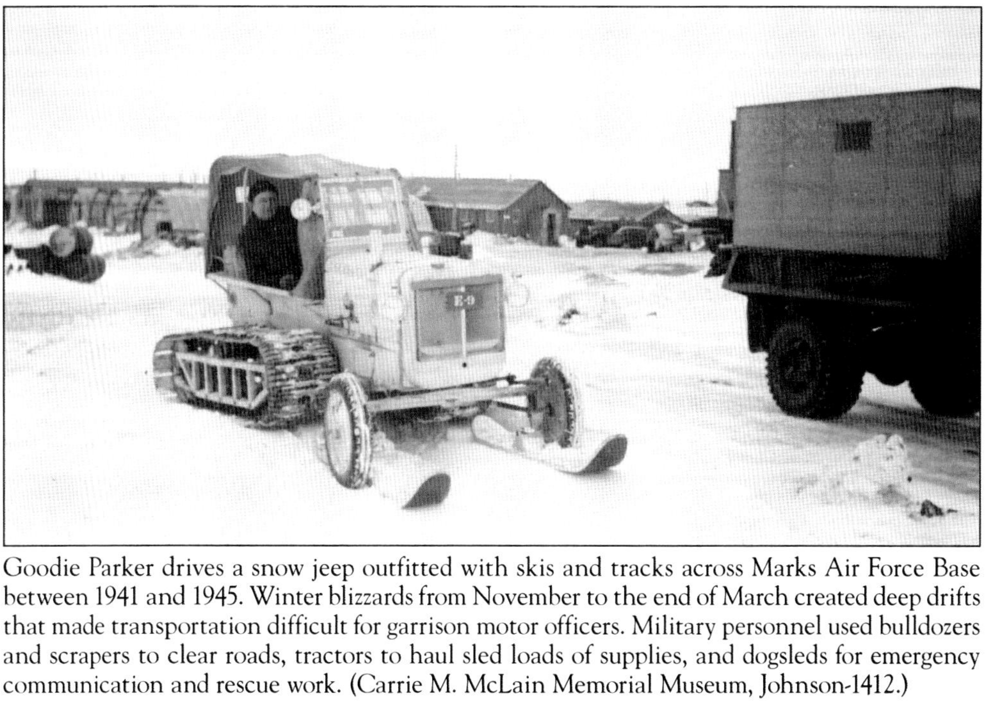

Goodie Parker drives a snow jeep outfitted with skis and tracks across Marks Air Force Base between 1941 and 1945. Winter blizzards from November to the end of March created deep drifts that made transportation difficult for garrison motor officers. Military personnel used bulldozers and scrapers to clear roads, tractors to haul sled loads of supplies, and dogsleds for emergency communication and rescue work. (Carrie M. McLain Memorial Museum, Johnson-1412.)

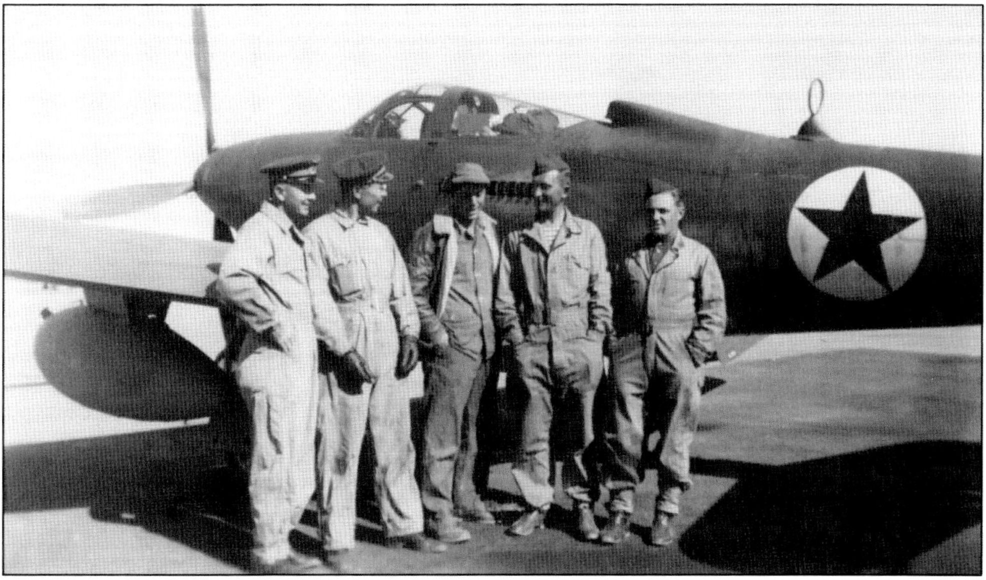

Russian and US military personnel pose in front of a Bell P-39 Airacobra used by the Soviet Air Force in World War II. Marks Air Force Base played an important role in the Lend-Lease Program, which supplied Allied countries with US military ships, planes, and supplies. Over 4,500 tactical aircraft were ferried from Nome to Siberia between 1942 and 1944. (Carrie M. McLain Memorial Museum, Lewis-2065.)

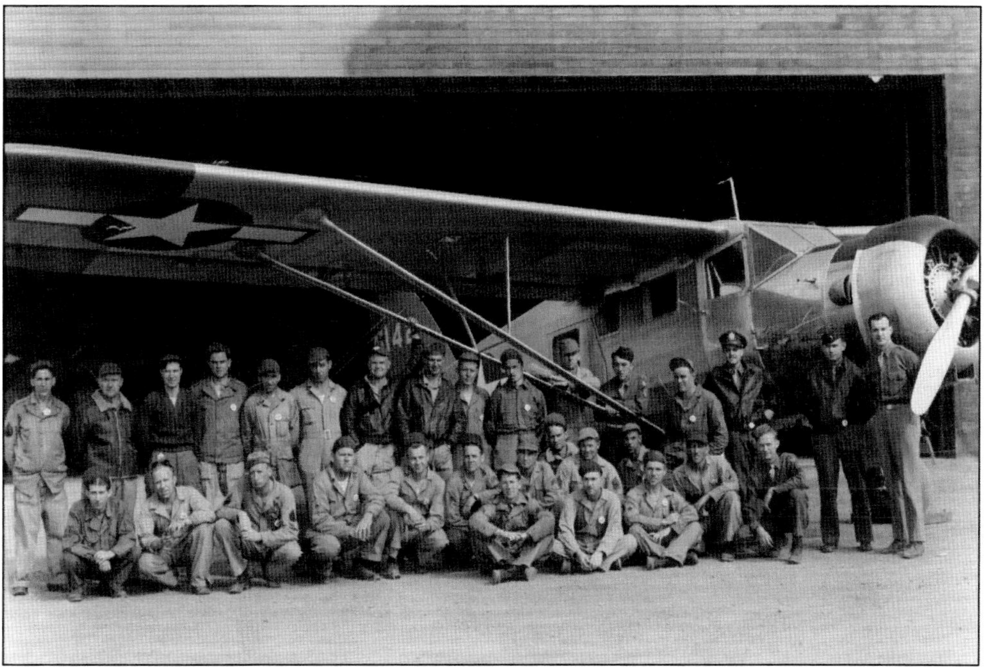

Military personnel gather in the doorway of a "T" hangar next to a UC-64 Norseman rebuilt by the 368th Service Squadron in Nome around 1943–1945. The "T" hangar was adjacent to a satellite field at Moonlight Springs that was completed in 1943. The satellite field served as a secondary official post with housing and administrative buildings. (Carrie M. McLain Memorial Museum, Storzek-1846.)

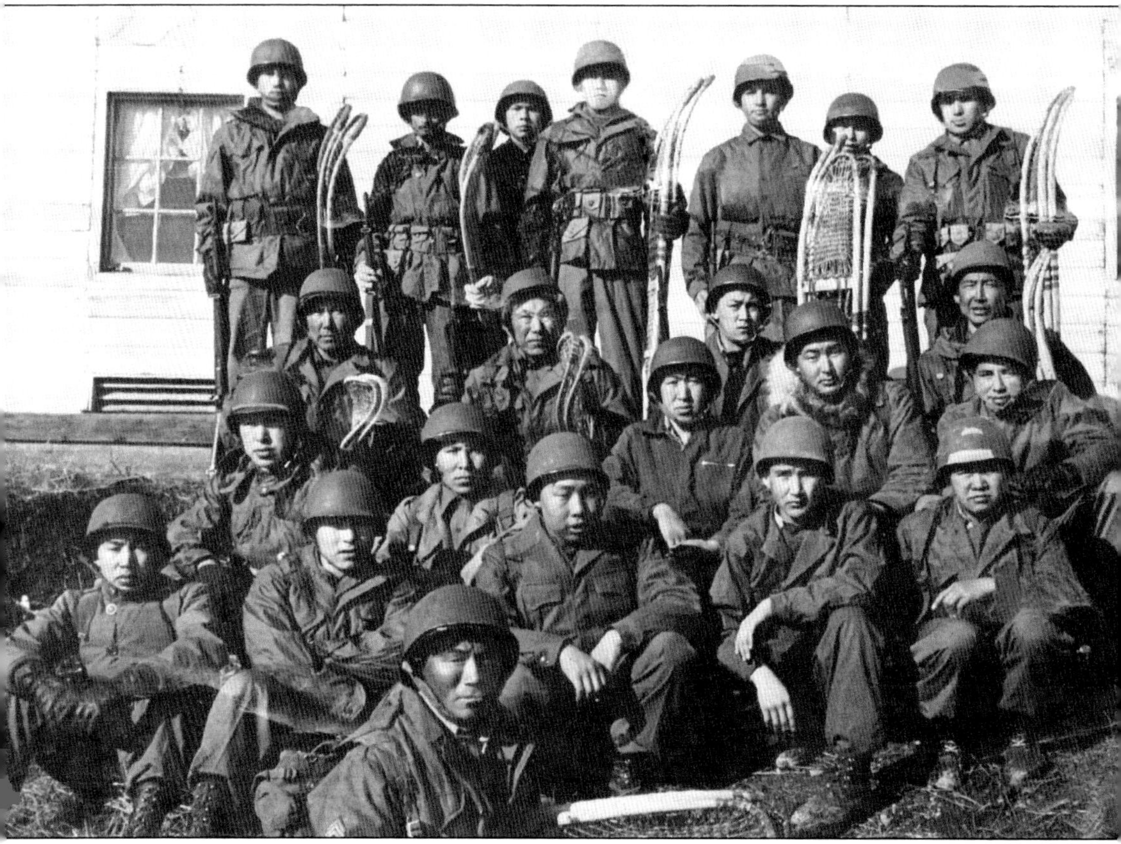

Members of the First Scout Battalion, Alaska Territorial Guard Unit of White Mountain gather for a group photograph around 1946. Pictured are, from left to right, (first row) William Oksoktaruk; (second row) two unidentified, Edwin Punguk, George Ashenfelter, and Mike Simon; (third row) Noah Napouk, Tom Brown Sr., David Henry, Nick Gray, and unidentified; (fourth row) Francis Walker, Edward Walker, William Soonagruk, and Steve Agloinga; (fifth row) unidentified, Wilson Shoogukwruk, and three unidentified; (sixth row) two unidentified. The Alaska Territorial Guard was a largely Alaska Native militia called up to protect the immense territory of Alaska from the threat of Japanese aggression during World War II. The 6,600 member unit was organized by Army Air Corps major Marvin "Muktuk Marston," who was based in Nome and traveled across the tundra by dog sled to deliver passionate speeches and sign up recruits. The guard was disbanded in 1947, but its success provided a model for the creation of the National Guard. (Carrie M. McLain Memorial Museum.)

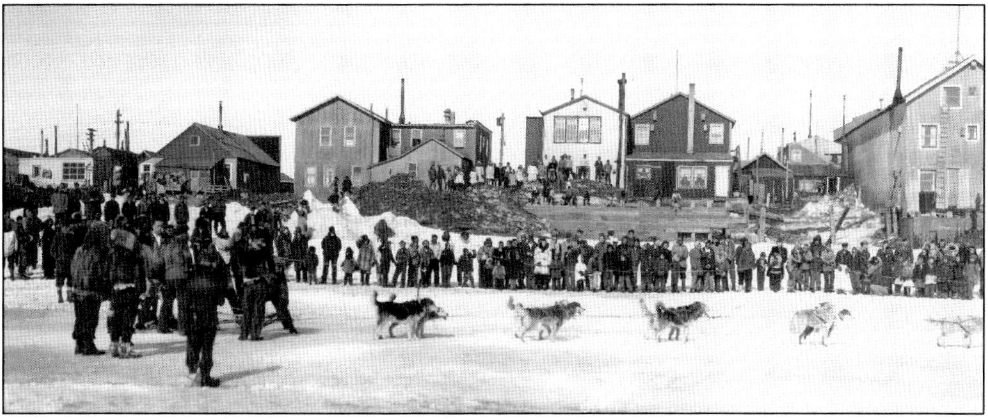

Nome residents and personnel from Marks Air Force Base wait anxiously for the start of a sled dog race on the Bering Sea ice. Military personnel held cross-country ski races, hunted Arctic hares, and tried their hand at mushing dogs during the long winters. Summer recreation included baseball tournaments, fishing for salmon and trout, and excursions to Pilgrim Hot Springs. (Carrie M. McLain Memorial Museum, Johnson-1112.)

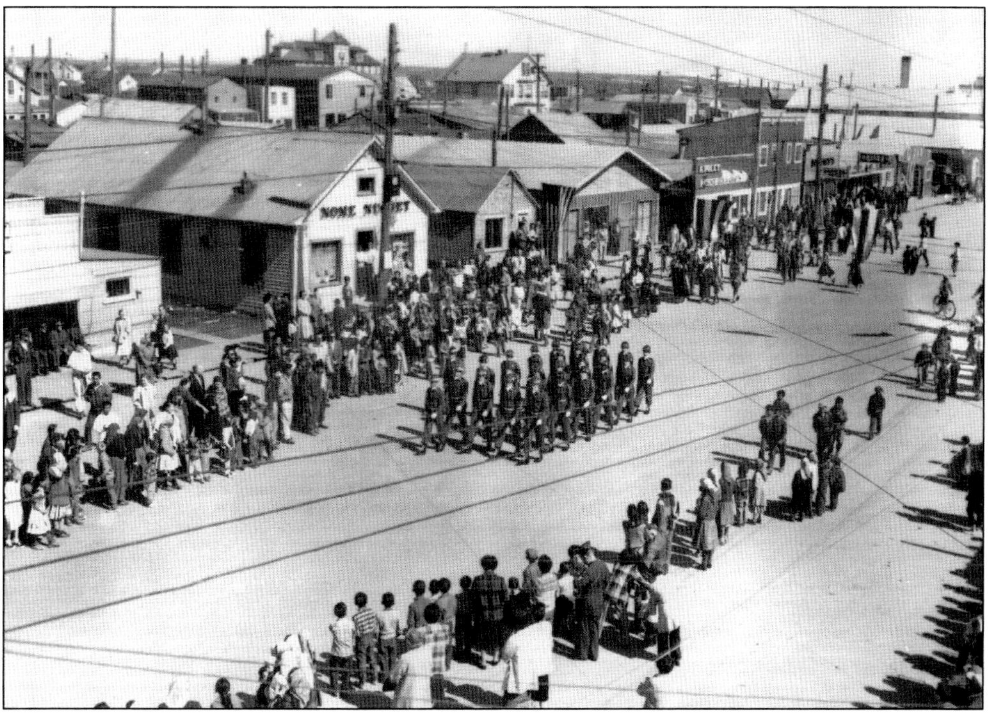

Crowds gather to watch members of the US Air Force march in a parade down Front Street during the 1950s. Marks Air Force Base was renamed Nome Field in 1951 with military activities focused on intelligence gathering during the Cold War era. Servicemen continue to be remembered and honored in Nome today with an annual memorial parade that extends from the Veterans of Foreign Wars post, down Front Street, and to the bank of the Snake River. (Carrie M. McLain Memorial Museum, NMP-15-31.)

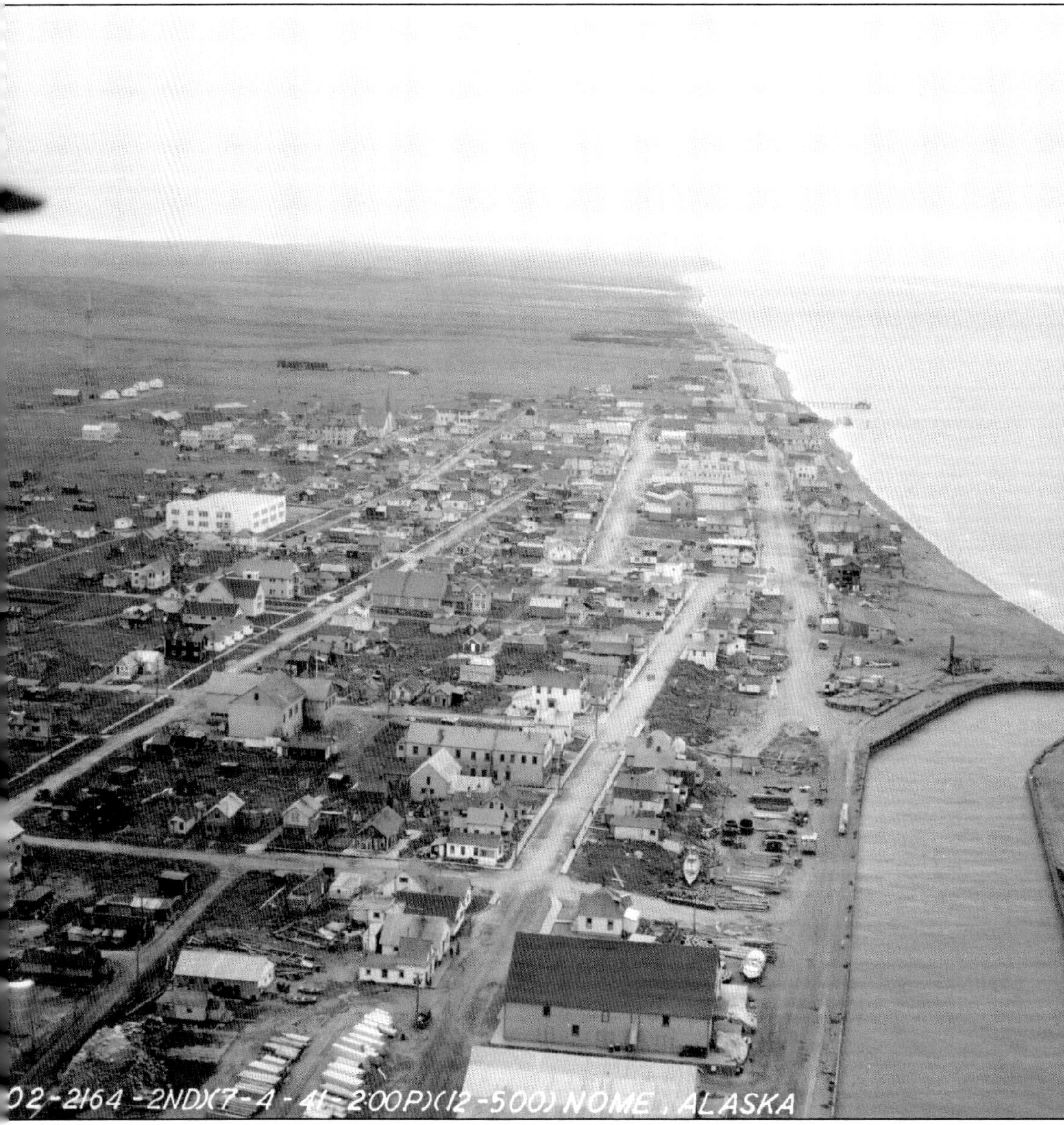

A military aircraft from Marks Air Force Base captured this view of Nome looking east over the Snake River on July 4, 1941. Buildings from different eras crowd next to one another along with boats, stacks of driftwood, and dog kennels. The new US post office and courthouse dominates the townscape on the left with its flat roof and smooth concrete walls. At right, the derelict piers and abandoned train in the distance speak to the increasing reliance on air transportation. The military airfield was decommissioned in 1956, and residents moved many of the Quonsets and panel buildings into town where they were used as storage, work spaces, and housing. (Carrie M. McLain Memorial Museum, 2006.15.34.)

BIBLIOGRAPHY

Burch, Ernest. *Social Life in Northwest Alaska: The Structure of Iñupiaq Eskimo Nations.* Fairbanks, AK: University of Alaska Press, 2006.

Cole, Terrence. *Nome: City of the Golden Beaches.* Anchorage, AK: Alaska Geographic, 1984.

Gillette, Gary. *Historic Buildings of Nome: Past/Present/Future.* Anchorage, AK: Bureau of Indian Affairs, Alaska Region, Division of Environmental and Cultural Resource Management, 2008.

Harrison, Edward S. *Nome and the Seward Peninsula.* Seattle, WA: Metropolitan Press, 1905.

Krug, John W. and Carol Sale Krug. *One Dog Short: The Odyssey and Collection of a Family in Alaska during the Gold Rush Years.* Juneau, AK: Division of Libraries, Archives, and Museums, Alaska Department of Education, 1998.

Kaplan, Lawrence, compiler. *Dictionary of King Island Inupiaq.* Fairbanks, AK: University of Alaska, Alaska Native Language Center, 1994.

Lewis, W. "US Army Report of Construction of Marks Field." May 1944. National Archives, Record Group 547.1 Records of Alaska Defense Command/Alaska Department, 1940–1947.

Mack, Sean, Rita Miraglia, and Cindy Williams. *Legacy of the Banner Creek Railroad Station.* Anchorage, AK: Bureau of Indian Affairs, Alaska Region, Division of Environmental and Cultural Resource Management, 2015.

McLain, Carrie M. *Gold-Rush Nome.* Portland, OR: Graphic Arts Center, 1969.

Nelson, Edward W. *The Eskimo about Bering Strait.* Washington, DC: Government Printing Office, 1899.

Ray, Dorothy Jean. *A Legacy of Arctic Art.* Seattle, WA: University of Washington Press, 1996.

Schrader, Frank C. and Alfred H. Brooks. *Preliminary Report on the Cape Nome Gold Region, Alaska.* Washington, DC: Government Printing Office, 1900.

Sutton, Anne and Sue Steinacher. *Alaska's Nome Area Wildlife Viewing Guide: Exploring the Nome Roadways.* Juneau, AK: Alaska Department of Fish and Game, 2012.

Thomas, Bob and Pam. *Leonhard Seppala: The Siberian Dog and the Golden Age of Sled Dog Racing 1908–1941.* Missoula, MT: Pictorial Histories Publishing, 2015

VanStone, James W., John A. Kakaruk, and Charles V. Lucier. "Reindeer Fairs on Seward Peninsula, Alaska 1915–1918." *Arctic Anthropology* 37(2): 60-77, 2000.